INTRODUCTION TO PROGRAMMABLE LOGIC CONTROLLERS

Lab Manual

INTRODUCTION TO PROGRAMMABLE LOGIC CONTROLLERS

Lab Manual

Gary A. Dunning

Delmar Publishers

an International Thomson Publishing company I(T)P®

Albany • Bonn • Boston • Cincinnati • Detroit • London • Madrid
Melbourne • Mexico City • New York • Pacific Grove • Paris • San Francisco
Singapore • Tokyo • Toronto • Washington

NOTICE TO THE READER

Cover Design:
Charles Cummings Advertising/Art, Inc.

Delmar Staff
Publisher: Alar Elken
Acquisitions Editor: Mark Huth
Art and Design Coordinator: Michele Canfield
Editorial Assistant: Dawn Daugherty

COPYRIGHT © 1998
By Delmar Publishers
an International Thomson Publishing Company

The ITP logo is a trademark under license.

Printed in the United States of America

For more information, contact:

Delmar Publishers
3 Circle, Box 15015
Albany, New York 12212-5015

International Thomson Publishing Europe
Berkshire House
168-173 High Holborn
London, WC1V 7AA
England

Thomas Nelson Australia
102 Dodds Street
South Melbourne, 3205
Victoria, Australia

Nelson Canada
1120 Birchmount Road
Scarborough, Ontario
Canada, M1K 5G4

International Thomson Editores
Campos Eliseos 385, Piso 7
Col Polanco
11560 Mexico D F Mexico

International Thomson Publishing GmbH
Konigswinterer Strasse 418
53227 Bonn
Germany

International Thomson Publishing Asia
221 Henderson Road
#05-10 Henderson Building
Singapore 0315

International Thomson Publishing--Japan
Hirakawacho Kyowa Building, 3F
2-2-1 Hirakawacho
Chiyoda-ku, Tokyo 102
Japan

Online Services

Delmar Online
To access a wide variety of Delmar products and services on the World Wide Web, point your browser to:
> http://www.delmar.com
> or email: info@delmar.com

A service of I(T)P

1 2 3 4 5 6 7 8 9 XXX 03 02 01 00 99 98

ISBN: 0-8273-7867-x

Library of Congress Catalog Card Number: 97-39868

TABLE OF

CONTENTS

PREFACE

The surest way to learn to do something is in doing it. The best way to learn how PLCs work, how they program, and how to edit programs is to get hands-on experience and configure, program and edit PLC Programs yourself. This lab manual is intended to accompany the *Introduction to Programmable Logic Controllers* text book. After you have studied the first nine introductory chapters in the text book, you are ready to try developing programs yourself. The programs you develop will be using the Allen-Bradley SLC 500 modular PLC, or the SLC 500 Fixed PLC, or the MicroLogix 1000 micro controller. Lab 2 introduces each of the three PLC configurations and how they are addressed. Even though lab programs are developed with a default modular SLC 500 with a 5/03 processor and a seven-slot chassis containing selected I/O modules, after understanding how the three PLCs are programmed, and how to convert the addresses from one to the other, modifying your lab exercise for your PLC, whether a fixed SLC 500 or MicroLogix, should pose no problem. If you have access to all three hardware configurations of the SLC family, you can develop programs with all three.

This lab manual will guide you through developing programs for the SLC 500, or MicroLogix 1000 micro controller using an IBM compatible personal computer and a recent version of Rockwell Software's APS software. Even though version 6.04 was used to develop these programming exercises, any recent version of APS should prove acceptable. Lab exercises will step you through I/O configuration, developing your first ladder program, parallel, latching, one-shot, various timers and counter programs, editing programs, printing hard copy programs, and adding documentation to selected programs. You will attach to your SLC 500 or MicroLogix, go on line, download and run most of your newly developed programs. You will learn about faulting the PLC processor as the result of an I/O configuration change, determining what the fault is using fault codes, and looking at the status file. You will also learn to recover from that fault once you determine the cause and make necessary corrections.

We will develop a *Begin* default processor file. You will save this on your personal floppy disk to use when developing upcoming programs. When you develop a new program you will load the *Begin* file and then rename it to reflect the current program. When completed, you will resave the new processor file to your floppy.

NOTICE TO LAB MANUAL USERS

The illustrations, programs and examples in this lab manual, along with its associated classroom activities, are intended solely to illustrate programming features of the Allen-Bradley programmable logic controllers mentioned. The author and others involved with the instruction of this subject cannot assume responsibility for the use of material contained herein for real applications. Examples contained herein, or mentioned or illustrated in class, have been simplified for instructional purposes. As a result of being simplified, these examples may not prove acceptable or safe in actual industrial applications. Always consult applicable electrical codes prior to any application with electrical equipment.

LAB MANUAL EXERCISE CORRELATION TO TEXT CHAPTERS

The following table illustrates the lab manual exercises that are suggested to correlate with text book chapters.

Chapter Number	Chapter Title	Suggested Lab Exercise
1	(Introduction) Welcome . . .	Lab Exercise 1
2	Number Systems	
3	Introduction to PLC Operation	Lab Exercise 2
4	Input Modules	
5	Output Modules	Lab Exercise 3
6	Putting Together a Modular PLC	Lab Exercise 4
7	Introduction to Logic	Lab Exercise 5
8	Programming a . . . Controller	
9	PLC Processors	
10	. . . Data Organization inside a PLC . . .	Lab Exercises 6 and 7
11	The Basic Relay Instructions	Lab Exercises 8, 9, 10, and 11
12	. . . Relay Instructions and the PLC	Lab Exercises 12 through 16
13	Documenting Your System	Lab Exercises 17, 18, and 19
14	Timer and Counter Instructions	Lab Exercises 20 through 29
15	Comparison and Data Handling . . .	Lab Exercises 30 through 35
16	Sequencer Instructions	Lab Exercises 36 through 39

INTRODUCTION TO PROGRAMMING

Lab exercises in this manual assume you have probably not programmed an SLC 500 using a personal computer and software before. Since this an introduction to programming a SLC 500 PLC, this manual will show you how to program basic ladder diagrams. We will not cover all PLC features, instructions, or system configurations. Our purpose is to provide your first hands-on programming exercises and PLC interface experience.

As of this writing there are five SLC 500 modular processors, the fixed SLC 500, and the MicroLogix 1000. There are also many ways to configure your system software along with many ways to connect your personal computer to the particular PLC you are working with. Your personal computer operating system might be DOS, Windows 3.1, Windows for Workgroups, Windows 95, or Windows NT. Your interface could be a direct serial connection using a null-modem cable, a

1747 PIC, PCMCIA Card, connection through a Data-Highway 485 network, or connection through a Data-Highway Plus network. There are a number of interface cards installable in your desktop personal computer's expansion slots. Possible interface cards include a 1784-KT, 1784-KTX, 1784-KTXD. Refer to your text for information on interface between a personal computer and PLC.

This lab manual makes the following assumptions:

1. Your instructor has Allen-Bradley APS software loaded on your personal computer. We will be using APS version 6.04 software.
2. Your instructor has demonstrated which method of personal computer to PLC interface is preferred in your lab setup.
3. Your SLC 500 hardware is either an SLC 5/02 or 5/03 modular processor in any of the available chassis, a fixed SLC 500, or a MicroLogix 1000.

Conventions

Introductory lab exercises will step you through each key you need to press to develop the program. Most information you will enter by pressing a series of function keys. Each step will have a number along with a check off space to help you track which steps you have completed. The steps below represent a sample of the steps you may execute as part of a Lab.

1. _____ Start from main screen from your APS software.
2. _____ Insert your floppy disk in drive A.
3. _____ Press F7, *File Options*.
4. _____ To copy from floppy select F8, *Copy from disk*.
5. _____ Select F5, *All of the above*, to copy entire file to your hard drive.
6. _____ Use the arrow keys to highlight the *Begin* file from the A drive directory.
7. _____ Press F3, *Select File*.
8. _____ Press F1, *Begin Operation*.

In most cases, after you have entered the introductory program, you will be asked to develop a similar program on your own. There will be review questions at the end of some lab exercises. We have tried to provide adequate answer space wherever needed, but you may need to furnish scratch paper of your own from time to time, especially when you are asked to make a drawing.

1

INTRODUCTION TO SLC PROGRAMMING

The SLC 500 uses the following basic bit or relay logic instructions to represent inputs and outputs. These instructions operate on a single bit of data associated with the bit's assigned address. Refer to Chapter 3, Bit Instructions, in the APS Reference Manual. Figure 1-1 illustrates the three basic bit instructions used in the SLC 500 and MicroLogix 1000 PLCs.

	Name	Symbol	Similar to	Processor Action	When True
SLC 500 BIT INSTRUCTIONS					
—-│ │—--	Examine if closed	XIC	Normally Open	Processor examines to see if closed.	Closed = True
—-│/│—--	Examine if open	XIO	Normally Closed	Processor examines to see if open.	Open = True
—( )—	Output Energize	OTE	Relay coil, or output	Proocessor turns rung output on when rung is true.	When rung true

Figure 1-1

Figure 1-2 illustrates SLC 500 and MicroLogix 1000 bit instructions symbology, name, and normal state, along with when each instruction is either true or false.

	Name	Normal State	
SLC 500 BIT INSTRUCTIONS			
—│ │—-	Examine if closed	Open	Nonenergized = FALSE Energized = TRUE
—│/│—-	Examine if open	Closed	Nonenergized = TRUE Energized = FALSE
—( )—	Output Energize	Off	Nonenergized = FALSE Energized = TRUE

Figure 1-2   The basic SLC 500 programming instructions.

UNDERSTANDING XIO AND XIC INSTRUCTIONS

During the processor scan, while running the user program, the processor examines the ON / OFF states of data file bits during the input update portion of the scan. After all inputs have been read and the input image table updated, the processor solves the ladder logic. During the program scan, one rung is scanned at a time by evaluating instructions starting from the leftmost on the current rung and working right in search of true instructions which will provide logical continuity. Logical continuity makes a rung true. As a result of a logically true rung, outputs are turned on during the output update portion of the scan.

To better understand logical continuity, let's compare it to electrical continuity. Electrical continuity is a complete path for electron flow through a circuit so as to turn on the controlled device. Figure 1-3 illustrates that for this conventional circuit to turn on light # 1, we must have current flow through switch #1 and switch #2. We now have a complete circuit, or electrical continuity, which will allow light # 1 to light.

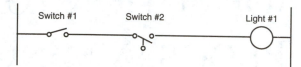

Figure 1-3 Electrical continuity needed to provide complete circuit.

Logical continuity in PLC ladder rungs operate in a similar fashion to electrical continuity. PLC ladder rungs, such as illustrated in Figure 1-4, do not have electrical continuity as in a conventional circuit. The PLC ladder rung illustrated in Figure 1-4 must have both instructions true before the processor will send an output signal to the output module. If inputs I:1 and I:2 are true, output L1 is true.

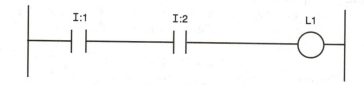

Figure 1-4 PLC ladder rungs have logical continuity, not electrical continuity.

Remember, input signals are placed in the appropriate data file during the input portion of the scan. When the logic is solved, during the program scan portion of the processor scan, the processor is looking for logical continuity between switch one and switch two. Each instruction is evaluated as either true or false. A true instruction is represented as a logical 1 whereas a false instruction is represented by a logical 0. These ones and zeros are placed in memory as a logical representation of the true or false state of each instruction. Logical continuity simply means that there is a continuous flow of "true" instructions from the left power rail to the output instruction. The processor then turns on or off the physical output devices as directed by the solved rung logical status placed in the output status file. Figure 1-5 illustrates the truth table for Figure 1-4's PLC ladder rung.

TRUTH TABLE FOR FIGURE 1-4 LOGIC		
Inputs		Outputs
XIC	XIC	OTE
False (0)	False (0)	False (0)
False (0)	True (1)	False (0)
True (1)	False (0)	False (0)
True (1)	True (1)	True (1)

Figure 1-5

CONTROLLING BITS OF LOGIC IN A LADDER PROGRAM

A ladder program consists of a number of individual rungs containing one or more input instructions, and usually one (however, there may be multiple) output instruction. The following exercise refers to the following two rungs of logic (Figure 1-6):

```
Rung 2:0
        I:1     I:1       I:1                                                            O:2
 -+--] [---] [--+--] [--[------------------------------------------------------------( )--
  |      0      5  |      7                                                             3
  |    I:1         |
  +--] [-----------+
         2

Rung 2:1
        I:1                                                                   O:2
 -+--] [------------+----------------------------------------------------+--( )--+-
  |       1         |                                                     |   4   |
  |     I:1    I:1  |                                                     |  O:2  |
  +--] [--+--] [--+-+                                                     +--( )--+
     2    |    6                                                             5
          |  I:1  |
          +--] [--+
              4
```

Figure 1-6   Two rungs of PLC ladder logic to fill in for Figure 1-7.

Identify each instruction type and list its address in the table in Figure 1-7. Instruction one is leftmost on rung one. Work from left to right and then to the next rung.

	Instruction	Address
# 1:		
# 2:		
# 3:		
# 4:		
#5:		
#6:		
#7:		
#8:		
#9:		
#10:		
#11:		

Figure 1-7   Instruction identification exercise for ladder runs from Figure 1-6.

Fill in the table in Figure 1-8 with information for the basic bit instructions. For additional information refer to your text, or Chapter 3, Bit Instructions, in the APS Instruction Set Reference Manual.

BASIC BIT INSTRUCTIONS DATA						
	XIO	XIC	OTL	OTU	OSR	OTE
Symbol						
Explanation						
When logic 0						
When logic 1						
Input or output?						

Figure 1-8   Table of basic bit instructions.

PROGRAM DEVELOPMENT

What makes up a PLC program? A PLC program and the associated data are called a *processor file*. A processor file consists of *program files* and *data files*. Each of these and the kinds of information it contains are listed below. As you work through the lab manual you will be working with many of the parts that make up these files.

Program Files

Program files are contained in a processor file. Program files contain the following information vital to processor operation.

1. System Information:
   A. I/O Configuration
   B. Processor Type
   C. Password
   D. Program Name (Processor File Name)
2. Main Ladder Program.
3. Subroutine Ladder Programs or Files.

Data Files

Data files are also part of a processor file. Data files are used as a storage locations for all data associated with instructions along with I/O points in the ladder program.

File 0. Input Status
File 1. Output Status
File 2. Processor Status Bits
File 3. Bit Storage
File 4. Timer
File 5. Counter
File 6. Control File Information
File 7. Integer Storage

2

SLC 500 ADDRESSING

Soon you will develop your first PLC ladder diagram using a personal computer and ladder development software. Since there are many SLC 500 processors, fixed SLC 500 PLCs and the MicroLogix PLC, lab exercises in this manual will be as generic as possible, so any of the available SLC family of PLCs can be used.

All labs were written for the default SLC 500 modular PLC, which we will introduce in the "If you are using a modular SLC 500 PLC" section. However, all exercises can be easily completed with either a MicroLogix 1000 PLC, or a fixed SLC 500 PLC. Listed below are the I/O specifications and addressing for either the MicroLogix 1000, or the fixed SLC 500 PLC. You can easily convert the addresses for the MicroLogix 1000 or fixed SLC 500 using the information below.

IF YOU ARE USING A MICROLOGIX 1000 PLC

The MicroLogix 1000 is a small footprint, fixed I/O PLC available in two configurations: ten inputs and six outputs, and twenty inputs and twelve outputs. MicroLogix 1000 I/O is currently not expandable beyond its current I/O count.

Manuals for the MicroLogix 1000 show the following addresses on the units themselves:

Sixteen-Point MicroLogix 1000:
O/0, O/1, O/2, O/3, O/4, O/5
I/0, I/1, I/2, I/3, I/4, I/5, I/6, I/7, I/8, I/9

Thirty-two-Point MicroLogix 1000:
O/0, O/1, O/2, O/3, O/4, O/5, O/6, O/7, O/8, O/9, O/10, O/11
I/0, I/1, I/2, I/3, I/4, I/5, I/6, I/7, I/8, I/9, I/10, I/11, I/12, I/13, I/14, I/15, I/16, I/17, I/18, I/19.

I/O PROGRAMMING ADDRESSES

As you develop user ladder programs, the correct syntax must be used whenever entering input and output addresses. If you attempt to enter incorrect address syntax, the APS software will give you an error message.

The proper addressing syntax for the MicroLogix 1000 is as follows:

Input zero	I:0/0		Output zero	O:0/0
Input one	I:0/1		Output one	O:0/1
Input two	I:0/2		Output two	O:0/2
Input three	I:0/3		Output three	O:0/3
Input four	I:0/4		Output four	O:0/4
Input five	I:0/5		Output five	O:0/5
Input six	I:0/6		Output six	O:0/6
Input seven	I:0/7		Output seven	O:0/7
Input eight	I:0/8		Output eight	O:0/8
Input nine	I:0/9		Output nine	O:0/9
Input ten	I:0/10		Output ten	O:0/10
Input eleven	I:0/11		Output eleven	O:0/11
Input twelve	I:0/12			
Input thirteen	I:0/13			
Input fourteen	I:0/14			
Input fifteen	I:0/15			
Input sixteen	I:0/16			
Input seventeen	I:0/17			
Input eighteen	I:0/18			
Input nineteen	I:0/19			

Attempting to configure any other I/O address—even using the correct format—which is not a normal, valid address for a MicroLogix 1000, will result in an "I/O address not configured" message. Keep in mind that a properly formatted I/O address which would be valid for a larger SLC 500 PLC can be programmed in a MicroLogix 1000 ladder program, even though the address may not be a valid MicroLogix 1000 address. Addressing errors will have to be corrected before the software will allow you to download and go on line.

MICROLOGIX 1000 I/O ADDRESSING FORMAT

The proper addressing format for a MicroLogix 1000 PLC is presented in Figure 2-1. MicroLogix 1000 part numbers and I/O configurations are listed below:

1761-L16AWA

120 / 240 VAC line power.
Ten 120 VAC inputs, addresses: I:0/0 through I:0/9.
Six relay outputs, addresses: O:0/0 through O:0/5.

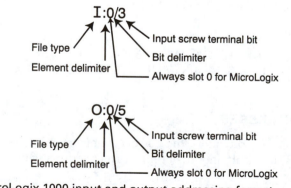

Figure 2-1   MicroLogix 1000 input and output addressing format.

1761-L32AWA

> 120 / 240 VAC line power.
> Twenty 120 VAC inputs, addresses: I:0/0 through I:0/19.
> Twelve relay outputs, addresses: O:0/0 through O:0/11.

1761-L32AAA

> 120 / 240 VAC line power.
> Twenty 120 VAC inputs, addresses: I:0/0 through I:0/19
> Two relay outputs, addresses: O:0/0 through O:0/1.
> Ten 120, 240 VAC triac outputs, addresses: O:0/2 through O:0/11.

1761-L16BWA

> 120 / 240 VAC line power.
> Ten 24 VDC inputs, addresses: I:0/0 through I:0/9
> Six relay outputs, addresses: O:0/0 through O:0/5

1761-L32BWA

> 120 / 240 VAC line power.
> Twenty 24 VDC inputs, addresses: I:0/0 through I:0/19.
> Twelve relay outputs, addresses: O:0/0 through O:0/11.

1761-L16 BWB

> 24 VDC line power
> Ten 24 VDC inputs, addresses: I:0/0 through I:0/9.
> Six relay outputs, addresses: O:0/0 through O:0/5.

1761-L32BWB

> 24 VDC line power
> Twenty 24 VDC inputs, addresses: I:0/0 through I:0/19.
> Twelve relay outputs, addresses: O:0/0 through O:0/11.

1761-L16BBB

> 24 VDC line power.
> Ten 24 VDC inputs, addresses: I:0/0 through I:0/9.
> Six DC MOSFET outputs, addresses: O:0/0 through O:0/5.

1761-L32BBB

> 24 VDC line power.
> Twenty 24 VDC inputs, addresses: I:0/0 through I:0/19.
> Twelve DC MOSFET outputs, addresses: O:0/0 through O:0/11.

IF YOU ARE USING A FIXED SLC 500 PLC

The fixed SLC 500 was the original fixed member of the SLC family, before the development of the MicroLogix 1000. The fixed SLC contains the processor, power supply, and I/O built into a single unit, just like the MicroLogix 1000. All I/O is built in, or fixed. The number of, and signal type of, I/O in a fixed PLC is not changeable. If a fixed PLC was ordered with twenty 120 VAC input points, that voltage level and number of inputs are fixed at the factory. The end user cannot change the I/O type or count.

SLC 500 fixed I/O PLC controllers come in three I/O configurations, 20, 30, or 40 I/O points. One optional two-slot expansion chassis can be clipped to the right end of the unit to add up to two additional *modular* I/O modules.

Using the optional two-slot expansion slot chassis, two I/O modules may be inserted in the expansion chassis, increasing the possible I/O count by 64 I/O. The expansion chassis can include input or output modules up to thirty-two points each. Specialty modules, like analog input or output, may reside in the expansion chassis.

SLC 500 Fixed PLCs are available as listed below (note that the last part of the part number indicates the total of that PLC's inputs and outputs).

1747-L20

Twelve inputs, addresses: I:0/0 through I:0/11.
Eight outputs, addresses: O:0/0 through O:0/7.

1747-L30

Eighteen inputs, addresses: I:0/0 through I:0/17.
Twelve outputs, addresses: O:0/0 through O:0/11.

1747-L40

Twenty-four inputs, addresses: I:0/0 through I:0/23.
Sixteen outputs, addresses: I:0/0 through I:0/15.

The table below (Figure 2-2) identifies the letter and the corresponding I/O configuration:

Part Number	Inputs	Outputs	Line Voltage
1747-L__A	120 VAC	Relay	120/240 VAC
1747-L__B	120 VAC	Triac	120/240 VAC
1747-L__C	DC Sink	Relay	120/240 VAC
1747-L__D	DC Sink	Triac	120/240 VAC
1747-L__E	DC Sink	DC Source	120/240 VAC
1747-L__F	DC Sink	Relay	24 VDC
1747-L__G	DC Sink	DC Source	24 VDC
1747-L__L	DC Source	DC Sink	120/240 VAC
1747-L__N	DC Source	DC Sink	24 VDC
1747-L__P	240 VAC	Triac	120/240 VAC
1747-L__R	240 VAC	Relay	120/240 VAC

Figure 2-2   SLC 500 fixed PLC part numbers and I/O correlation.

I/O Programming Addresses

As you develop user ladder programs, the proper syntax must be used whenever entering input and output addresses. If you attempt to enter incorrect address syntax, the APS software will give you the same error message as the MicroLogix 1000. Addressing a fixed SLC 500 also follows the same format as the MicroLogix 1000.

IF YOU ARE USING A MODULAR SLC 500 PLC

When complete flexibility in I/O count and I/O mix, processor power features and memory size are desired, a modular PLC is required for the application. A modular SLC 500 can be specifically configured to fit almost any application. Being modular, the processor, power supply, chassis and required I/O modules are individually selected from the many selections available.

Chassis are used to hold all the selected pieces together. Chassis are selected by the number of *slots* available for module insertion. Typically any module can be inserted into any slot. Slots are numbered from left to right starting at slot zero. The processor always goes in slot zero. Even though the processor is in slot zero, a modular PLC, unlike a fixed PLC, has no I/O associated with slot zero. All chassis start with slot zero as the leftmost slot in the chassis. SLC 500 chassis slot numbers are listed below:

A four-slot chassis has slots 0 through 3.
A seven-slot chassis contains slots 0 through 6.
The ten-slot chassis consists of slots 0 through 9.
Thirteen-slot chassis have slots 0 through 12.

Figure 2-3 below illustrates SLC chassis and slot assignments.

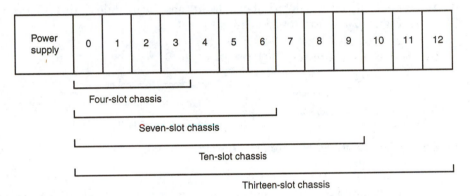

Figure 2-3   SLC 500 chassis slot numbering assignments.

The modular SLC 500 PLC may contain up to three chassis connected together. Any mix of the available chassis may be connected to increase the total I/O count; a total of thirty I/O slots is allowed in any modular SLC 500 PLC. Figure 2-4 is an example of a three-chassis SLC 500 PLC.

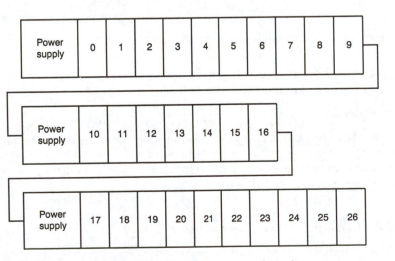

Figure 2-4   Three chassis expanded locally to provide a total of 27 slots.

Determining a Discrete I/O Address

The processor must be able to differentiate one I/O module from another in each chassis. Each I/O point on each module also must be uniquely identified.

In order to uniquely identify a module type, slot, and I/O point, an *address* is constructed using all of this information. The module type, O for output and I for input, the chassis slot number, and the module's individual I/O screw terminal data are used to construct the input or output point address. This I/O address is the identifying data you will associate with each instruction on a ladder program.

Each chassis slot may be either an input or an output. If using combination input and output modules, a single slot can contain input as well as output addresses. Each I/O point must have a unique address which is reflected in either the input data file, or the output data file. The addressing format is very similar to the MicroLogix or fixed SLC 500 PLCs. The only difference between fixed I/O addressing and modular I/O addressing is the slot number will always be zero for fixed PLCs whereas the slot number for addressing I/O points for a modular SLC 500 PLC will be any valid slot value from one to thirty—never zero. The addressing format for outputs consists of module type (O for output), chassis slot number (1 to 30), and the module's individual output screw terminal (0 to 31). Figure 2-5 illustrates addressing format.

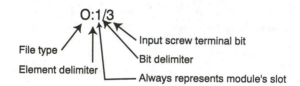

Figure 2-5   Modular SLC 500 output addressing format.

The addressing format for inputs is very similar to output addressing. Figure 2-6 illustrates input addressing format.

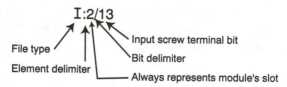

Figure 2-6   Modular SLC 500 input addressing format.

Figure 2-7 can be used to convert modular PLC input and output addresses used for exercises in this manual to either fixed SLC 500 addresses, or MicroLogix addresses.

CONVERTING MODULAR I/O ADDRESSING TO FIXED I/O ADDRESSING			
Modular Input Address	**Fixed Input Address**	**Modular Output Address**	**Fixed Output Address**
I:1/0	I:0/0	O:1/0	0:0/0
I:1/1	I:0/1	O:1/1	0:0/1
I:1/2	I:0/2	O:1/2	0:0/2
I:1/3	I:0/3	O:1/3	0:0/3
I:1/4	I:0/4	O:1/4	0:0/4
I:1/5	I:0/5	O:1/5	0:0/5
I:1/6	I:0/6	O:1/6	0:0/6

Figure 2-7   Table for converting modular to fixed SLC 500 I/O addresses. *(continues)*

CONVERTING MODULAR I/O ADDRESSING TO FIXED I/O ADDRESSING			
Modular Input Address	**Fixed Input Address**	**Modular Output Address**	**Fixed Output Address**
I:1/7	I:0/7	O:1/7	O:0/7
I:1/8	I:0/8	O:1/8	O:0/8
I:1/9	I:0/9	O:1/9	O:0/9
I:1/10	I:0/10	O:1/10	O:0/10
I:1/11	I:0/11	O:1/11	O:0/11
I:1/12	I:0/12	O:1/12	O:0/12
I:1/13	I:0/13	O:1/13	O:0/13
I:1/14	I:0/14	O:1/14	O:0/14
I:1/15	I:0/15	O:1/15	O:0/15

Figure 2-7   Table for converting modular to fixed SLC 500 I/O addresses. *(continued)*

3

FILE AND ADDRESS WORKSHEET

1. Complete the address format information in Figure 3-1.

Input or Output?	Chassis Slot?	Screw Terminal?	Address
I	1	4	I:1/4
Output	2	?	O:2/13
Input	?	1	I:4/1
Input	12	11	?
?	11	?	O:11/9
Output	3	3	?
Input	?	?	I:14/5
Input	9	15	?
Input	1	?	I:1/8
?	?	?	O:13/15
Output	5	?	O:5/9
Input	6	0	?

Figure 3-1 SLC 500 addressing exercise.

2. Illustrated in Figure 3-2 is a seven-slot SLC 500 chassis. Fill in the appropriate bit positions in the associated input and output files with the appropriate status bits for each module. The module in slot one, an input module, has been completed as a guide.

Power supply	Slot 0 CPU	Slot1 Input	Slot 2 Input	Slot 3 Output	Slot 4 Input	Slot 5 Output	Slot 6 Output

Figure 3-2 Modular SLC 500 PLC representation.

The module in slot one has screw terminals 0, 1, 7, 9 and 14 as ON conditions.
The module in slot two has screw terminals 2, 3, 5, 9 and 10 as ON conditions.
The module in slot three has screw terminals 0, 1, 3 and 15 as ON conditions.
The module in slot four has screw terminals 2, 7, 8, 9 and 14 as ON conditions.
The module in slot five has screw terminals 1, 7, 9, 11 and 14 as ON conditions.
The module in slot six has screw terminals 14 and 15 as ON conditions.

	0	1	2	3	4	5	6	7	8	9	10	11	12	13	14	15
0:3																
0:5																
0:6																

Figure 3-3   Output status file for the PLC represented in Figure 3-2.

	0	1	2	3	4	5	6	7	8	9	10	11	12	13	14	15
I:1	1	1	0	0	0	0	0	1	0	1	0	0	0	0	1	0
I:2																
I:4																

Figure 3-4   Input status file for the PLC represented in Figure 3-2.

3. What is a word?_____

4. How many bits make up an SLC 500 memory word?_____

5. Define bit._____

6. Define file._____

7. What is a processor file?_____

8. What is a program file?_____

9. What is the purpose of a data file?_____

10. In which other file is the data file located?_____

11. How many processor files may reside in the CPU at any one time?_____

12. How many program files can be contained in a processor file?_____

13. How many data files can be contained in a processor file?_____
14. Identify what is contained in program files:

   File 0:_____

   File 1:_____

   File 2:_____

   File 3:_____
15. Identify what is contained in each data file in Figure 3-5 on the next page.

16. Are all SLC address formats the same? Why, or why not?_____

File number	Identifier	File type
0		
1		
2		
3	B	
4		
5		Counter
6		
7		
8		
9		

Figure 3-5   SLC 500 data file.

17. What are the three main items that make up an I/O address?_____

_____

18. What data is contained in the input data files?_____

    How is individual data represented?_____

    Where does this data come from?_____

    What does this data represent?_____

19. What are the three things that make up an input data file?_____

_____

One of these is used to represent input signals. How does it do that?_____

_____

20. What are the three things that make up an output data file?_____

_____

One of these is used to represent output signals. How does it do that?_____

_____

4

ALLEN-BRADLEY SLC 500 POWER CALCULATION

This lab exercise will give you practice selecting the correct power supply as part of assembling a list of hardware components into a working Allen-Bradley SLC 500 PLC.

To configure an Allen-Bradley SLC 500 modular PLC, first determine the application's hardware needs. You need to select the specific chassis, power supply, CPU and I/O modules required for your particular application. All these modular pieces are assembled in the selected chassis to build the PLC for your specific application. Part of the selection process includes calculating the power required to operate the selected SLC 500 hardware components.

PREVIOUS KNOWLEDGE REQUIRED TO SUCCESSFULLY COMPLETE THIS LAB

Review examples on calculating power supply loading for the Allen-Bradley SLC 500, Chapter 6 of *Introduction to Programmable Logic Controllers*.

ADDITIONAL MATERIALS NEEDED

For this exercise you will need:

1. A calculator.
2. Power calculation exercise contained in this lab exercise.
3. Reference information from I/O module power tables from Chapter 6 in your text.

THE LAB

Figure 4-1 (next page) illustrates the modular SLC 500 PLC we are going to configure for this exercise. Determining the proper power supply consists of:

1. Listing the processor along with all modules to be installed in the chassis.
2. Fill in the minimum 5-volt and 24-volt loading current for each device in the chassis in a table similar to Figure 4-2. Data for completing the table can be found in Chapter 6 of your text.
3. Total each column to obtain the total 5-volt and 24-volt loading currents.
4. Select proper power supply.

	CPU	IN	IN	IN	OUT	OUT	OUT
?? Power Supply	5/03	IB16	OB16	OB16	IB8	OB16	OX8

Figure 4-1 SLC 500 modular PLC for this exercise.

Determine the power supply for the following seven-slot SLC 500 PLC. Fill in Figure 4-2 and complete the power loading calculations. Refer to your text and SLC 500 reference manuals for data.

SLC 500 MODULAR POWER CALCULATIONS: RACK #			
Slot	Hardware Description and Catalog Number	5VDC Loading Current	24 VDC Loading Current
0	CPU # 1747- (5/03)		
1	1746-IB16		
2	1746-OB16		
3	1746-OB16		
4	1746-IB8		
5	1746-OB16		
6	1746-OX8		
	TOTALS:		

Figure 4-2 SLC 500 power calculations exercise.

QUESTIONS

1. What power supply did you choose for chassis?_____

2. Explain why you chose this power supply._____

   _____

3. If you added a handheld programming terminal to your SLC 500 would this affect your current calculations?_____

4. What if anything would happen if the power supply loading was excessive for an SLC 500 PLC system?_____

SUMMARY

When configuring a PLC system, the user must take into consideration the load each module, CPU, programmer or interface converter to your personal computer will put on the rack or chassis power supply. Also, consideration must be taken regarding any user power taken out of the chassis power supply for peripheral devices. When calculating load on an SLC 500 power supply, 5-VDC along with 24-VDC loading current must be verified for being within specifications. These calculations are for the worst case. That is, the total load on the power supply is the load when all input and output points are energized on each module in your chassis.

Each power supply is designed to handle a specific load. In the event the power supply in any PLC rack or chassis should become overloaded, unpredictable or inconsistent system operation or power supply shutdown could occur.

5

DEVELOPING LADDER LOGIC

I/O is a common term used to represent input and output. In any control situation, there are controlling signals coming into a control panel while decision signals go out of the panel to the controlled device. These incoming signals are called *inputs*, while outgoing decision signals are called *outputs*.

The first task that needs to be completed before you can develop a PLC user program is either to develop a ladder diagram that is compatible with PLC symbology, or upgrade a current conventional schematic so it can be used to develop a PLC user program.

1. Determine Real-World I/O and Allocate Addresses.

    When converting a conventional ladder diagram schematic to one in which we can develop our user program, the first task is to determine inputs and outputs that are connected to real-world hardware devices. After real-world I/O have been determined, allocate each I/O point a valid input or output address.

2. Internal Reference Allocation.

    Once the real-world I/O references have been determined, internal coil instruction references need to be allocated. These internal coil instruction references are analogous to relay system relays that do not drive real-world devices but rather interact with other internal system relays. Internal coils are ladder diagram logic elements that interact with other non–real-world input or output logic internally in the CPU.

    Most PLCs allocate a portion of data memory for the storage of internal I/O statuses. If your particular PLC does not have internal coil references, you can assign unused actual output references for internal use.

3. Develop Ladder Logic in PLC Format.

4. Develop User Program.

5. Documentation of the User Program.

    A table should be developed that defines each input, output and internal coil reference. This table should list each I/O's function. Drawings that indicate the wiring and its operation should be included for future reference. The tables should list every point available for use, even if it be for a future use. A table should be prepared for internal PLC data storage addresses. This table should be filled in during the development of the program. Documentation of your user program will be covered in the section on documentation.

DEVELOPING LADDER RUNGS FROM FUNCTIONAL SPECIFICATIONS

Before you can develop successful ladder programs you must be able to translate functional specifications into PLC ladder rungs. As an example, take the following specification:

When push button PB2 is pressed, pilot light PL2A will turn on.

Now develop a rung of PLC ladder logic from this specification. Figure 5-1 illustrates a rung of PLC logic where the push button is represented as a normally open contact (XIC) instruction and the output pilot light is represented by an output coil (OTE) instruction.

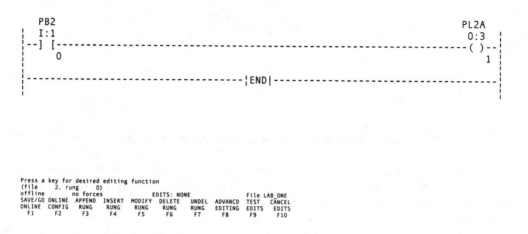

Figure 5-1 PLC ladder rung developed from functional specification.

For the PLC ladder diagram in Figure 5-2 complete the following:

1. List inputs and outputs:_____

2. Assign addresses to each input and output. Assume you have one sixteen-point input module with addresses I:1/0 through I:1/15 and one output module with addresses O:2/0 through

   O:2/15._____

```
:---] [---------------] [---] [---------------------------------( )---:
:   :               :   :   :                               :       :
:+-] [---] [---] [--+   +-]/[-+                           +-] [---( )-+:
:   :               :                                               :
:   +-] [---] [-+                                                    :

:---] [---] [---] [---] [---------------------------------( )---:
:   :                                                   :   :
:+-] [---]/[-+                                         +-( )-+:
```

Figure 5-2 Developing ladder rung from functional specifications.

For each of the following functional specifications develop the correct PLC ladder rung.

1. When push button PB2 is pressed, and switch SW2 is closed, pilot light PL2A will turn on.
2. When push button PB2 is pressed, or limit switch LS1 is closed, pilot light PL2A will turn on.
3. If inductive proximity switch SW1, and SW2, and SW3 all sense a target, motor M1 will start.
4. If any of the four doors of an automobile are open, the dome light will come on.
5. If limit switch SW1, and limit switch SW2 or limit switch SW3, are true, the full case is in position, energize glue gun to apply glue to case flaps.
6. If limit switches SW1 or SW2, and SW3, are true, energize solenoid SOL3A to move product into position.
7. If product is in position from the movement of SOL3A from question six above, energize outputs O:3/2 and O:3/7.
8. Input I:1/11 is a sensor that determines if there are box blanks in the feeder. If the boxes are not replenished by the operator and the feeder runs out of box blanks, the conveyor will be shut down and an alarm bell will sound.
9. A conveyor line is used to label and fill cans with product. A bar-code reader is used to read the bar code on the can's label to determine that the proper label has been placed on the can.

   A photoelectric sensor is used to trigger the bar-code reader when there is a can in position. The bar-code reader will then read the can's bar code. If the bar-code reader fails to see a label with a bar code or sees a bad or damaged bar code, a no-read discrete signal will be sent to the PLC. Develop two rungs of PLC logic: one rung for the bar-code read trigger, and a second rung to alert the PLC of a no-read condition.
10. A variable frequency drive has four preset speeds it can run at depending on the conditions of three inputs from a four-position selector switch into our PLC. The drive is an Allen-Bradley 1336 Plus Variable Frequency Drive with an interface card to accept 120-volt AC control signals. Input signal patterns into terminals 26, 27, and 28 determine at which preset speed the drive will run. The table in Figure 5-3 lists the conditions terminals 26, 27, and 28 need to be in to select a specific preset speed.

Preset Speed	Drive Option Card Input Signals for Speed Selection		
	26	27	28
Preset Speed 1	False	False	False
Preset Speed 2	False	False	True
Preset Speed 3	False	True	False
Preset Speed 4	True	True	False

Figure 5-3 Variable-speed drive preset speed input truth table.

Figure 5-4 illustrates the target table for the four-position selector switch.

Four Position Selector Switch Target Table				
Position and Preset #1	Position and Preset #2	Position and Preset #3	Position and Preset #4	Switch Circuit Input to PLC
0	0	0	X	1
0	0	X	0	2
0	X	0	0	3

Figure 5-4

Illustrate an overview of our application. Include the selector switch, input module, ladder logic, output module, and drive option card wiring to terminals 26, 27, and 28.

6

CREATING A DEFAULT PROCESSOR FILE

Each time a new SLC 500 or MicroLogix PLC ladder program is developed, a new processor file must be created. The processor file will be created using the APS software.

Included in the creation of the processor file is assigning a name to the file and entering data on the processor and the processor operation system, rack and configuration of the I/O modules in the rack(s).

This lab exercise will step you through the procedure to create a new processor file, enter processor and operating system data, configure the PLC's I/O, name the processor file *Begin*, and save the newly created processor file to your floppy disk. This floppy disk will be your beginning, or default, processor file for future programming exercises. Using this as a default processor file, you will not have to go through all the steps to create a new file each time you begin a new programming exercise.

MATERIALS YOU SHOULD HAVE AS YOU COMPLETE THESE EXERCISES

1. A copy of the Advance Programming Software (APS) User Manual and an SLC 500 Instruction Set Reference Manual should be available for your reference.
2. APS software should be running on your personal computer.

DEVELOPING A NEW PROCESSOR FILE

You will always begin at the main menu—the first screen that appears when you start up the APS software. Figure 6-1 (next page) is a picture of the APS Main Menu screen.

THE MAKE-UP OF THE MAIN MENU

The main menu is made of the following information. The top portion of the screen identifies:

1. The product as SLC 500 Advanced Programming Software.

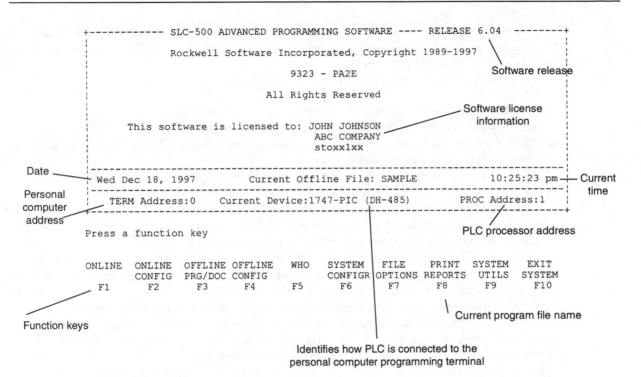

Figure 6-1 APS software main menu screen.

2. The software release as 6.04.
3. The software part number as 9323-PA2E.
4. Who the software is registered to, along with the software serial number.
5. The date, time, and current off-line file name.
6. The "Current Device" as a 1747-PIC (DH-485). If you are using the PIC as your interface between the personal computer and your PLC this will be your current device.

Personal Computer Address (TERM Address)

Your personal computer is considered the programming terminal. In order to communicate effectively with other devices, such as your PLC processor, each device must have a unique identifier called its address. Our programming device default address is 0. Terminal addresses can be changed by the user.

PLC Processor Address (PROC Address)

Like the terminal, the processor must have a unique address. For our example, the processor default address is 1. The processor address, like the terminal address, can also be assigned by the user. After a processor has been installed and communication with it verified, it should have its processor address changed to an address other than the default. The reason for this is that if you need to place another processor on your PLC network, its default address will be 1, and this node address will be available. Remember, each hardware device must have a unique address in order to communicate.

Function Keys

The APS programming software is a DOS-based programming package, rather than a Windows software package. As such, a mouse cannot be used as with Windows. Instead, the function keys are used to navigate around the software. The function keys on the computer screen correlate with the function keys on your personal computer keyboard. Each menu in the programming software will have its own specific function keys.

The main menu's function keys provide the following:

F1, On-line: The F1 selection allows you to go on-line with the PLC processor.

F2, On-Line Configuration: On-Line Configuration is where the on-line configuration of the programming terminal is set up. The programming terminal's port, terminal address, processor address, and communication baud rate are set up in this screen.

F3, Off-Line Programming and Development: Selecting *Off-Line Programming* provides you access to screens for developing a ladder program.

F4, Off-line Configuration: Selecting *Off-Line Configuration* allows you to select a processor file from the displayed directory, usually the hard drive, and go off-line and program or edit the file. In addition to this, a new processor file may be created, or the user directories for processor memory, comments, symbols, documentation reports and user configuration can be changed.

F5, Who: The SLC 500 modular, fixed and MicroLogix 1000 PLCs can be connected together on a network called the *Data Highway 485 Network (DH-485)*. Each PLC and programming terminal on the network must have a unique address so communication can be successful between a programming terminal and a selected PLC, or between one PLC and another PLC. There may be up to thirty devices on a single Data Highway network. Each device and its associated unique address is called a *node* on the network. Figure 6-2 illustrates an example of a DH-485 network. As illustrated in the example, this particular network consists of a programming terminal as node address 0, modular SLC 500 as node address 1, a fixed SLC 500 PLC as node address 2, and a MicroLogix 1000 as node address 4.

The *Who* selection allows you to view which addresses are in use on the Data Highway network. Node and network diagnostic windows may be viewed, programmer on-line and network configuration may be changed. Figure 6-2 is a printout of a *Who* screen illustrating our sample network.

F6, System Configuration: Your programming terminal system configuration is changed from this screen. Monitor color, printer configuration, and system startup display configuration are set up or modified here.

```
+----------- SLC-500 ADVANCED PROGRAMMING SOFTWARE ---- RELEASE 6.04  --------+
|                                                                             |
|                   Allen-Bradley Company, Copyright 1989-1997                |
|        +- WHO ACTIVE - Active Station Identification -----------------------+
|        |                                                                    |
|        |   0 TERM        (31) - 12              24                          |
|        |   1 5/03        (31) - 13 -> Maximum node      25                  |
|        |   2              14       address is set at 31 26                  |
|        |   3        Personal computer 15                27                  |
|        |   4        terminal, node 0  16                28                  |
|        |   5              17                    29                          |
|        |   6        SLC 500 5/03   18                30                     |
|        |   7        processor at node 1  19             31                  |
|        |   8              20                                                |
|        |   9              21                                                |
|        |  10              22                                                |
|        |  11              23                                                |
+----+- ESC exits ------------------------------------------------------------+

Press a key for desired network function


ONLINE   STATION   SET     CLEAR              MAX     NODE     NODE
         DIAGS     OWNER   OWNER            ADDRESS  ADDRESS   BAUD
   F1       F2      F3       F4                F7       F8      F9
```

Figure 6-2 Sample DH-485 network.

F7, File Options: The *File Options* selection is where you change the name of, copy, and delete processor files. Processor files can be copied from your personal computer's hard drive to a floppy, or from a floppy to the personal computer's hard drive. You will save your default processor file to your student floppy disk using this function.

F8, Print Reports: Hard copy reports are printed from this functions screen. Before a report can be printed, the report must be generated. Report generation is under another function screen on another screen.

F9, System Utilities: If you want to transfer a program to or from nonvolatile memory, or transfer a processor file between two terminals or between a terminal and a hand-held programming terminal.

F10, Exit System: The *Exit System* selection is used to exit the APS programming software and return you to DOS.

DEVELOPING A NEW PROCESSOR FILE

To create a new processor file, start at the main menu and complete the following steps.

1. _____ Press F3, *Off-Line Programming / Documentation.* Figure 6-3 illustrates your computer's screen.

```
+------------------- SLC-500 ADVANCED PROGRAMMING SOFTWARE -------[  OFFLINE  ]-+
|+- PROGRAM DIRECTORY FOR PROCESSOR: SAMPLE  -----------------------------------+
||  FILE   PROTECTED        NAME         TYPE              SIZE (words)         |
||   0                                   system                *                |
||   1                                   reserved              *                |
||   2      Yes                          ladder                *                |
||                                                                              |
||                                                                              |
||                                                                              |
||                                                                              |
||                                                                              |
||                                                                              |
||                                                                              |
||                                                                              |
||                                                                              |
||                                                                              |
||                                                                              |
++-----------------------------------------------------------------------------+

Press a key, enter file number or file name

offline                 SLC 5/02                              File SAMPLE
PROCSSR   SAVE    RETURN  CHANGE        CREATE   FILE   MONITOR  DATA   MEMORY
FUNCTNS           TO MENU FILE          REPORTS  OPTIONS FILE   MONITOR  MAP
  F1      F2       F3      F4             F6       F7     F8      F9      F10
```

Figure 6-3 Your screen after pressing F3, *Off line Programming.*

2. _____ Press F4, *Change File.*
   Figure 6-4 (next page) illustrates a sample directory of processor files currently on your hard drive. For this exercise we will be creating a new file rather than selecting a file from the list.

3. _____ Press F6, *Create File.*

4. _____ Type the processor file name *Begin.*

5. _____ Press *Enter.*
   Figure 6-5 (next page) illustrates your computer screen after the processor file name has been entered. The currently displayed processor may not be the one we will use for this program. We will enter the processor data next.

```
+----------------- SLC-500 ADVANCED PROGRAMMING SOFTWARE -------[ OFFLINE ]-+
|+- PROGRAM DIRECTORY FOR PROCESSOR: SAMPLE  -----------------------------------+
||   FILE   PROTECTED     NAME        + C:\IPDS\ARCH\SLC500 ----------------+
||    0                                | Name        Size      Date        |
||    1                                |-----------------------------------|
||    2        Yes                     |                                   |
||                                     | LAB_ONE     929     12-18-97       |
||                                     | SAMPLE      813     12-18-97       |
||                                     |                                   |
||                                     |                                   |
||                                     |                                   |
||                                     |                                   |
||                                     |                                   |
||                                     |                                   |
||                                     |                                   |
||                                     |                                   |
++-----------------------------------+- ESC exits/Alt-U aborts changes -----+
NEW ARCHIVE FILE CREATED
Press a Function Key or Enter File Name

offline                  SLC 5/03                            File SAMPLE
OFFLINE                                                      SAVE
PRG/DOC                              CREATE  DEFINE          TO FILE
 F1                                   FILE    DIR             F9
                                       F6      F7
```

Figure 6-4 Computer screen after pressing F4, *Change file.*

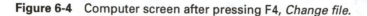

```
+------------------ SLC-500 ADVANCED PROGRAMMING SOFTWARE -------[ OFFLINE ]-+
|+- PROGRAM DIRECTORY FOR PROCESSOR: BEGIN  -----------------------------------+
||+- PROCESSOR -- INPUTS ----- OUTPUTS ---------+\ARCH\SLC500 ----------------+
|||  Bul. 1761   MicroLogix 1000                |        Size      Date       |
|||  1747-L511   5/01 CPU - 1K USER MEMORY       |----------------------------|
|||  1747-L514   5/01 CPU - 4K USER MEMORY      |NE      259    12-18-97      |
|||  1747-L524   5/02 CPU - 4K USER MEMORY      |E       813    12-18-97      |
|||  1747-L532   5/03 CPU -12K USER MEMORY      |                             |
|||  1747-L541   5/04 CPU -12K USER MEMORY      |                             |
||+--------------------------------------------+                             |
||                                         |                                  |
||+- CREATE PROCESSOR FILE ----------------------------------------+         |
|||                                                                |         |
|||   Name:        BEGIN                                           |         |
||| F2 Processor:  1747-L532 5/03 CPU - 12K USER MEMORY            |         |
|||                                                                |         |
|||                                                                |         |
+++- ESC exits/Alt-U aborts changes ------------------------------+changes -----+
Press a Function Key or Enter File Name
BEGIN
offline                  SLC 5/03                            File BEGIN
       SELECT                   CONFIGR ADJUST       SAVE &
       PROC                       I/O   FILTERS      EXIT
        F2                        F5      F6          F8
```

Figure 6-5 Computer screen after entering processor file name *Begin* and pressing *Enter.*

Default SLC 500 Hardware Used in This Manual

The default SLC 500 in a seven-slot chassis was used to develop the lab exercises for this lab manual, as listed in Figure 6-6 (next page).

The SLC 500 PLC that you will be using to complete these lab exercises is shown in Figure 6-7 (next page).

If you have a fixed SLC 500 or MicroLogix 1000 PLC, continue with the next section. If you have a modular SLC 500 skip to the 5/01 and 5/02 section, or the 5/03 and 5/04 section.

DEFAULT SLC 500 PLC USED TO DEVELOP LAB EXERCISES			
Slot	Module Part Number	I/O Points	Addresses
0	Processor SLC 5/03	None	None
1	1746-IB16	16 Inputs	I:1/0 - I:1/15
2	1746-OB16	16 Outputs	O:1/0 - O:1/15
3	1746-OB16	16 Outputs	O:1/0 - O:1/15
4	1746-IB8	8 Inputs	I:1/0 - I:1/7
5	1746-OB16	16 Outputs	O:1/0 - O:1/15
6	1746-OX8	8 Outputs	I:1/0 - I:1/7

Figure 6-6   Default SLC 500 PLC.

THE SLC 500 PLC I WILL USE TO DEVELOP LAB EXERCISES			
Slot	Module Part Number	I/O Points	Addresses
0			
1			
2			
3			
4			
5			
6			

Figure 6-7   My SLC 500 PLC.

If You Have a Fixed SLC 500 PLC or a MicroLogix 1000 PLC

I/O Configuration for an SLC 500 Fixed PLC, or MicroLogix 1000 PLC, is a manual process. The procedure below lists the steps to configure your fixed or MicroLogix PLC.

1. _____ User arrow keys to highlight and select the correct fixed SLC 500 part number, or the MicroLogix 1000 part number.
2. _____ Press *Enter.* Your screen should look similar to Figure 6-8.

```
+------------------- SLC-500 ADVANCED PROGRAMMING SOFTWARE -------[ OFFLINE ]-+
|+- PROGRAM DIRECTORY FOR PROCESSOR: BEGIN  ----------------------------------+
||+- PROCESSOR -- INPUTS ----- OUTPUTS ---------+\ARCH\SLC500 ----------------+
||| Bul. 1761   MicroLogix 1000                 |          Size       Date     |
||| 1747-L511   5/01 CPU - 1K USER MEMORY       |------------------------------|
||| 1747-L514   5/01 CPU - 4K USER MEMORY       |NE       259      12-18-97    |
||| 1747-L524   5/02 CPU - 4K USER MEMORY       |E        813      12-18-97    |
||| 1747-L532   5/03 CPU -12K USER MEMORY       |                              |
||| 1747-L541   5/04 CPU -12K USER MEMORY       |                              |
||+---------------------------------------------+                              |
||                                 |                                           |
|||+- CREATE PROCESSOR FILE -------------------------------------+            |
|||                                                              |            |
|||    Name:                                                     |            |
|||| F2 Processor:   Bul. 1761   MicroLogix 1000                 |            |
|||                                                              |            |
|||                                                              |            |
+++- ESC exits/Alt-U aborts changes ---------------------------+changes -----+

Press a Function Key or Enter File Name
BEGIN
offline                   1761 MicroLogix 1000                     File BEGIN
        SELECT                   CONFIGR ADJUST        SAVE &
        PROC                     I/O     FILTERS       EXIT
        F2                       F5      F6            F8
```

Figure 6-8   Computer screen after selecting a MicroLogix 1000 and naming the processor file.

```
+------------------- SLC-500 ADVANCED PROGRAMMING SOFTWARE -------[  OFFLINE  ]-+
|+- PROGRAM DIRECTORY FOR PROCESSOR: BEGIN  ------------------------------------+
|| FILE    PROTECTED        NAME     + C:\IPDS\ARCH\SLC500 ------------------+
|| 0                                 | Name          Size      Date          |
|| 1                                 |------------------------------------------|
|| 2       Yes                       | LAB_ONE       259       12-18-97       |
||                                   | SAMPLE        813       12-18-97       |
||                                   | BEGIN         259       12-18-97       |
||                                   |                                          |
||                                   |                                          |
||                                   |                                          |
||                                   |                                          |
||                                   |                                          |
||                                   |                                          |
||                                   |                                          |
||                                   |                                          |
||                                   |                                          |
|+------------------------------------+- ESC exits/Alt-U aborts changes -----+
++
NEW ARCHIVE FILE CREATED
Press a Function Key or Enter File Name

offline              1767 MicroLogix 1000                      File BEGIN
OFFLINE                                                        SAVE
PRG/DOC                               CREATE  DEFINE           TO FILE
  F1                                  FILE    DIR
                                      F6      F7               F9
```

Figure 6-9   Computer screen after new processor archive file has been created.

3. _____ Press F8, *Save and Exit.*
4. _____ You should see the message, "New Archive File Created," as illustrated in Figure 6-9.
5. _____ Press F1, *Off-Line Programming / Documentation.*
6. _____ F2, *Save.*
7. _____ F8, *Yes* to accept defaults.
8. _____ Save your processor file to your student floppy disk by selecting F7, *File Options.*
9. _____ F7, *Copy* to disk.
10. _____ F5, to select all of the above.
11. _____ Use the arrow keys to select the *Begin* file.
12. _____ F3, *Select File.*
13. _____ F1, *Begin Operation.* Your *Begin* file is saved to your student floppy disk in drive A.
14. _____ *Escape.*
15. _____ F3 to return to the main menu.
16. _____ Skip to Lab Exercise Seven.

If You Have an SLC 5/01 or 5/02 Processor

(If you have a 5/03 or 5/04 processor, skip to that section.) I/O Configuration for an SLC 5/01 or 5/02 processor is a manual process. The procedure below lists the steps to configure your PLC's I/O.

1. _____ Use the arrow keys to select your processor.
2. _____ Press F2, *Select Processor.* Figure 6-10 (next page) shows your computer screen after selecting a 5/02 processor.

Now that the processor has been selected, the next step is to configure the rack and I/O.

3. _____ Press F5, *Configure I/O.*
4. _____ To select your particular rack, Press F4, *Modify Racks.*
5. _____ There is only one rack in our default system, Press F1, *Select Rack One.*

```
+------------------- SLC-500 ADVANCED PROGRAMMING SOFTWARE -------[  OFFLINE ]-+
|+- PROGRAM DIRECTORY FOR PROCESSOR: BEGIN  ---------------------------------+
||+- PROCESSOR -- INPUTS ----- OUTPUTS ---------+                            |
|||  Bul. 1761    MicroLogix 1000               |                            |
|||  1747-L511    5/01 CPU - 1K USER MEMORY     |                            |
|||  1747-L514    5/01 CPU - 4K USER MEMORY     |                            |
|||  1747-L524    5/02 CPU - 4K USER MEMORY     |                            |
|||  1747-L532    5/03 CPU -12K USER MEMORY     |                            |
|||  1747-L541    5/04 CPU -12K USER MEMORY     |                            |
||+---------------------------------------------+                            |
||                                                                           |
||+- CHANGE PROCESSOR FILE -----------------------------------+              |
|||                                                           |              |
|||    Name:        BEGIN                                     |              |
||| F2 Processor:   1747-L524    5/02 CPU - 4K USER MEMORY    |              |
|||                                                           |              |
|||                                                           |              |
+++- ESC exits/Alt-U aborts changes ---------------------------+------------+

Press a Function key, ENTER to Select Processor, ESC to exit, or Alt-U to abort

offline                                                         File BEGIN
          SELECT                    CONFIGR ADJUST        SAVE &
          PROC                        I/O   FILTERS       EXIT
           F2                         F5      F6           F8
```

Figure 6-10 Computer screen after processor selection.

6. _____ Use the up or down arrow keys to highlight the rack you will be using.
7. _____ With the correct rack highlighted, press *Enter.*
8. _____ This returns you to the I/O configuration screen, similar to Figure 6-11.

```
+------------------- SLC-500 ADVANCED PROGRAMMING SOFTWARE -------[  OFFLINE ]-+
|+- PROGRAM DIRECTORY FOR PROCESSOR: BEGIN  ---------------------------------+
||+- I/O CONFIGURATION FOR:BEGIN  -------------------------------------------+
|||  RACK 1 =        1746-A47    7 slot Backplane                           |
|||  RACK 2 =        NOT INSTALLED                                          |
|||  RACK 3 =        NOT INSTALLED                                          |
|||                                                                         |
|||    SLOT        CATALOG #     CARD DESCRIPTION                           |
|||  *  0          1747-L524   5/02 CPU - 4K USER MEMORY                    |
|||  *  1                                                                   |
|||  *  2                                                                   |
|||  *  3                                                                   |
|||  *  4                                                                   |
|||  *  5                                                                   |
|||  *  6                                                                   |
|||     7                                                                   |
|||     8                                                                   |
+++- ESC exits -------------------------------------------------------------+

Press a function key

offline                                                         File BEGIN
  READ      ONLINE          MODIFY  MODIFY  DELETE   UNDEL   EXIT   SPIO
  CONFIG    CONFIG          RACKS   SLOT    SLOT     SLOT           CONFIG
   F1        F2              F4      F5      F6       F7     F8      F9
```

Figure 6-11 Computer screen after rack selection.

I/O Module Configuration

Notice there are stars to the left of the valid slot numbers in your selected rack. These stars identify how many slots need to have modules assigned. To configure your I/O modules:

1. _____ Slot one should be highlighted. If it is not highlighted, use the up or down arrows to highlight this slot.
2. _____ Press F5, *Modify Slot.*
3. _____ Using the up or down arrows, from the list, select the module that is installed in slot one. Figure 6-12 illustrates the I/O module selection screen.

```
+------------------- SLC-500 ADVANCED PROGRAMMING SOFTWARE -------[  OFFLINE ]-+
|+- PROGRAM DIRECTORY FOR PROCESSOR: BEGIN    -------------------------------+
||+- I/O CONFIGURATION FOR:BEGIN      ---------------------------------------+
|||+- I/O MODULE SELECTION FOR SLOT:1  ------------------+
||||   CATALOG #     CARD DESCRIPTION                    |
||||   1746-I*8   Any  8pt. Discrete Input Module        |
||||   1746-I*16  Any 16pt. Discrete Input Module        |
||||   1746-I*32  Any 32pt. Discrete Input Module        |
||||   1746-O*8   Any  8pt. Discrete Output Module       |
||||   1746-O*16  Any 16pt. Discrete Output Module       |
||||   1746-O*32  Any 32pt. Discrete Output Module       |
||||   1746-IA4    4-Input 100/120 VAC                   |
||||   1746-IA8    8-Input 100/120 VAC                   |
||||   1746-IA16  16-Input 100/120 VAC                   |
||||   1746-IB8    8-Input (SINK) 24 VDC                 |
||||   1746-IB16  16-Input (SINK) 24 VDC                 |
||||   1746-IB32  32-Input (SINK) 24 VDC                 |
++++- ESC exits -----------------------------------------+---------------+

Press ENTER to select I/O Module
Enter Module ID Code>
offline                                              File BEGIN
        SELECT
        MODULE
          F2
```

Figure 6-12   I/O module selection screen.

4. _____ When the module has been highlighted, press F2 to select the module.
5. _____ You will be returned to the I/O configuration screen. The selected module should be displayed in slot one similar to Figure 6-13 (next page).
6. _____ To add the I/O module for slot two, press the down arrow to highlight slot two.
7. _____ Press F5 to add a module to slot two.
8. _____ Use the arrow keys to highlight the I/O module residing in slot two.
9. _____ When the correct module is selected, press F2, or *Enter.*
10. _____ You will be returned to the I/O configuration screen. The selected module should be displayed in slot two. See Figure 6-14 (next page).
11. _____ Press the down arrow to select slot three. Go through the steps to select the module for slot three. Figure 6-15 (page 30) illustrates your computer screen after selection of I/O module in slot three.
12. _____ Continue selecting I/O modules in this manner until all I/O modules have been selected.
13. _____ When all I/O modules have been selected, press F8, *Exit.*
14. _____ Press F8 to save the I/O configuration and exit.
15. _____ This will return you to a screen listing all program files currently on your computer's hard drive. The newly developed program file named *Begin* should be highlighted.

```
+------------------- SLC-500 ADVANCED PROGRAMMING SOFTWARE -------[ OFFLINE ]-+
|+- PROGRAM DIRECTORY FOR PROCESSOR: BEGIN    ----------------------------------+
||+- I/O CONFIGURATION FOR:BEGIN    ------------------------------------------+
|||  RACK 1 =       1746-A7   7-slot Backplane                               |
|||  RACK 2 =       NOT INSTALLED                                            |
|||  RACK 3 =       NOT INSTALLED                                            |
|||                                                                          |
|||    SLOT         CATALOG #      CARD DESCRIPTION                          |
|||   *  0          1747-L524   5/02 CPU - 4K USER MEMORY                    |
|||   *  1          1746-IB16   16-Input (SINK) 24 VDC                       |
|||   *  2                                                                   |
|||   *  3                                                                   |
|||   *  4                                                                   |
|||   *  5                                                                   |
|||   *  6                                                                   |
|||      7                                                                   |
|||      8                                                                   |
+++- ESC exits --------------------------------------------------------------+

Press a function key

offline                                                        File BEGIN
  READ    ONLINE          MODIFY  MODIFY  DELETE   UNDEL   EXIT  SPIO
 CONFIG   CONFIG          RACKS   SLOT    SLOT     SLOT          CONFIG
   F1       F2             F4      F5      F6       F7      F8     F9
```

Figure 6-13   Selected I/O module displayed in slot one of our rack.

```
+------------------- SLC-500 ADVANCED PROGRAMMING SOFTWARE -------[ OFFLINE ]-+
|+- PROGRAM DIRECTORY FOR PROCESSOR: BEGIN    ----------------------------------+
||+- I/O CONFIGURATION FOR:BEGIN    ------------------------------------------+
|||  RACK 1 =       1746-A7   7-slot Backplane                               |
|||  RACK 2 =       NOT INSTALLED                                            |
|||  RACK 3 =       NOT INSTALLED                                            |
|||                                                                          |
|||    SLOT         CATALOG #      CARD DESCRIPTION                          |
|||   *  0          1747-L524   5/02 CPU - 4K USER MEMORY                    |
|||   *  1          1746-IB16   16-Input (SINK) 24 VDC                       |
|||   *  2          1746-OB16   16-Input (TRANS) (SOURCE) 10 / 50 VDC        |
|||   *  3                                                                   |
|||   *  4                                                                   |
|||   *  5                                                                   |
|||   *  6                                                                   |
|||      7                                                                   |
|||      8                                                                   |
+++- ESC exits --------------------------------------------------------------+

Press a function key

offline                                                        File BEGIN
  READ    ONLINE          MODIFY  MODIFY  DELETE   UNDEL   EXIT  SPIO
 CONFIG   CONFIG          RACKS   SLOT    SLOT     SLOT          CONFIG
   F1       F2             F4      F5      F6       F7      F8     F9
```

Figure 6-14   Selected I/O module displayed in slot two of our rack.

```
+------------------- SLC-500 ADVANCED PROGRAMMING SOFTWARE -------[ OFFLINE ]-+
|+- PROGRAM DIRECTORY FOR PROCESSOR: BEGIN   --------------------------------+
||+- I/O CONFIGURATION FOR:BEGIN   ------------------------------------------+
|||  RACK 1 =        1746-A7    7-slot Backplane                             |
|||  RACK 2 =        NOT INSTALLED                                           |
|||  RACK 3 =        NOT INSTALLED                                           |
|||                                                                          |
|||     SLOT        CATALOG #       CARD DESCRIPTION                         |
|||  *  0           1747-L524    5/02 CPU - 4K USER MEMORY                   |
|||  *  1           1746-IB16    16-Input (SINK) 24 VDC                      |
|||  *  2           1746-OB16    16-Input (TRANS) (SOURCE) 10 / 50 VDC       |
|||  *  3           1746-OB16    16-Input (TRANS) (SOURCE) 10 / 50 VDC       |
|||  *  4                                                                    |
|||  *  5                                                                    |
|||  *  6                                                                    |
|||     7                                                                    |
|||     8                                                                    |
++- ESC exits ---------------------------------------------------------------+

Press a function key

offline                                                          File BEGIN
  READ    ONLINE             MODIFY  MODIFY  DELETE  UNDEL  EXIT   SPIO
CONFIG    CONFIG             RACKS   SLOT    SLOT    SLOT          CONFIG
  F1        F2                 F4      F5      F6      F7    F8      F9
```

Figure 6-15 I/O module selected for chassis slot three.

16. _____ Press F9 to save the new configuration to file. This completes the I/O configuration process. Figure 6-16 illustrates a sample screen with the program file *Begin* included.

17. _____ If you wanted to develop a new ladder program at this point, you would press F1, *Off-Line Programming / Documentation*. We want to save this processor file to our student floppy disk as our default beginning file for future lab exercises.


```
+------------------- SLC-500 ADVANCED PROGRAMMING SOFTWARE -------[ OFFLINE ]-+
|+- PROGRAM DIRECTORY FOR PROCESSOR: BEGIN   --------------------------------+
||  FILE   PROTECTED      NAME    + C:\IPDS\ARCH\SLC500 ----------------+
||   0                            |  Name        Size       Date        |
||   1                            |-----------------------------------  |
||   2     Yes                    |  LAB_ONE      259     12-18-97       |
||                                |  SAMPLE       813     12-18-97       |
||                                |  BEGIN        259     12-18-97       |
||                                |                                     |
||                                |                                     |
||                                |                                     |
||                                |                                     |
||                                |                                     |
||                                |                                     |
||                                |                                     |
||                                |                                     |
||                                |                                     |
++-------------------------------+- ESC exits/Alt-U aborts changes -----+
NEW ARCHIVE FILE CREATED
Press a Function Key or Enter File Name

offline            SLC 5/02                                   File BEGIN
OFFLINE                          CREATE  DEFINE        SAVE
PRG/DOC                           FILE    DIR          TO FILE
  F1                               F6      F7            F9
```

Figure 6-16 Program directory listing the newly created file and named *Begin*.

```
+------------------- SLC-500 ADVANCED PROGRAMMING SOFTWARE -------[  OFFLINE ]-+
|+- PROGRAM DIRECTORY FOR PROCESSOR: BEGIN     ---- SINGLE STEP TEST: ENABLED  -+
||  FILE   PROTECTED      NAME        TYPE                    SIZE (words)   |
||   0                                system                  0             |
||   1                                reserved                0             |
||   2       No                       ladder                  5             |
||                                                                          |
||                           +- File Operations ---------+                  |
||                           |                           |                  |
||                           | F3 Rename                 |                  |
||                           | F4 Copy                   |                  |
||                           | F5 Delete                 |                  |
||                           |                           |                  |
||                           | F7 Copy To Disk           |                  |
||                           | F8 Copy From Disk         |                  |
||                           |                           |                  |
++-------------------------+- ESC exits -------------+----------------------+

Press a function key

offline            SLC 5/02                                File BEGIN
              RENAME    COPY    DELETE          COPY    COPY
                                             TO DISK  FR DISK
                F3       F4       F5            F7       F8
```

Figure 6-17 File operations menu.

18. _____ Press *Escape*.
19. _____ Press F7, *File Options*. Figure 6-17 illustrates the file operations menu.
20. _____ Press F7, *Copy to Disk*.
21. _____ Press F5 to select all options.
22. _____ Use the arrow keys to select the file to be copied to your student floppy disk. Make sure the *Begin* program is highlighted.
23. _____ Press F3 to select the *Begin* file.
24. _____ Press F1 to begin the operation of copying the *Begin* program from the hard disk to your student floppy.
25. _____ After a few seconds the floppy drive light will go off, and the operation is completed.
26. _____ *Escape*.
27. _____ *Escape*.
28. _____ Press F3 to return to the main menu.
29. _____ You have now created your default processor file named *Begin* and stored a copy on your student floppy disk. Keep this disk safe as it will be used to create and store lab programming exercises as you continue through this lab manual. Some ladder programs that you develop will be saved to your student floppy disk to be later retrieved, renamed and edited.
30. _____ Skip to the section, I/O Configuration Errors and Faulting Your Processor.

If You Have an SLC 5/03 or 5/04 Processor

If you have a 5/03 or 5/04 processor, the I/O configuration can be automatically configured by attaching your programming device to the processor and reading the I/O configuration. Have your personal computer attached to your SLC 500 PLC as demonstrated by your instructor. The procedure below lists the steps to perform an automatic configuration.

```
+------------------- SLC-500 ADVANCED PROGRAMMING SOFTWARE -------[ OFFLINE ]-+
|+- PROGRAM DIRECTORY FOR PROCESSOR: BEGIN   ---------------------------------+
||+- I/O CONFIGURATION FOR:BEGIN    ------------------------------------------+
|||  RACK 1 =        1746-A7    7-slot Backplane                              |
|||  RACK 2 =        NOT INSTALLED                                            |
|||  RACK 3 =        NOT INSTALLED                                            |
|||                                                                          |
|||    SLOT         CATALOG #       CARD DESCRIPTION                          |
|||   *  0          1747-L532     5/03 CPU -12K USER MEMORY                   |
|||   *  1                                                                    |
|||   *  2                                                                    |
|||   *  3                                                                    |
|||   *  4                                                                    |
|||   *  5                                                                    |
|||   *  6                                                                    |
|||      7                                                                    |
|||      8                                                                    |
++++- ESC exits ------------------------------------------------------------+

Press a function key

offline                    SLC 5/03                             File BEGIN
  READ    ONLINE           MODIFY  MODIFY  DELETE  UNDEL  EXIT    SPIO
CONFIG    CONFIG           RACKS    SLOT    SLOT   SLOT          CONFIG
  F1        F2               F4      F5      F6     F7     F8      F9
```

Figure 6-18 Your computer screen after selection of a 5/03 processor.

PART I

1. _____ Using the arrow keys, highlight the 5/03 or 5/04 processor.
2. _____ Press F2, *Select Processor.* Figure 6-18 illustrates your computer screen after selection of a 5/03 processor.
3. _____ A box will pop up for you to select your processor's operating system. Use the arrow keys to select your processor's operating system.

   If you are using a 5/03 processor your choices are:

   1747-OS300 Series A
   1747-OS301 Series A
   1747-OS302 Series A

   Figure 6-19 (next page) illustrates the screen with a 5/03 processor's operating system selections.

   If you are using a 5/04 processor your choices are:

   1747-OS400 Series A
   1747-OS401 Series A
   1747-OS402 Series A

   To find the operating system on your processor: The easiest way to determine the current operating system of your processor is to look at the left side of the processor, just below the key switch. The sticker should have the current O.S. # (operating system number).

   For more information on processor operating systems, refer to your text.
4. _____ Press *Enter.*
5. _____ Press F5, *Configure I/O.*
6. _____ Press F1, *Read I/O.* (Make sure you are attached to your PLC.)
7. _____ The message "Read I/O Configuration" should display.
8. _____ Press F8, *Yes.* The display should be filled in with all the PLC's I/O modules.

```
+------------------- SLC-500 ADVANCED PROGRAMMING SOFTWARE -------[ OFFLINE ]-+
|+- PROGRAM DIRECTORY FOR PROCESSOR: BEGIN    --------------------------------+
||+- PROCESSOR -- INPUTS ----- OUTPUTS ---------+\ARCH\SLC500 ----------------+
|||+------------ OPERATING SYSTEM -------------+|            Size      Date    |
|||  1747-OS300 SERIES A                       ||---------------------------- |
|||  1747-OS301 SERIES A                       |E          813     12-18-97   |
|||  1747-OS302 SERIES A                       |                              |
|||                                            |                              |
||+--------------------------------------------+                              |
||+--------------------------------------------+                              |
|||                                      |                                    |
|||+- CREATE PROCESSOR FILE -------------------------------------+            |
|||                                                              |            |
|||    Name:         BEGIN                                       |            |
||| F2 Processor:   1747-L532   5/03 CPU -12K USER MEMORY        |            |
|||                                                              |            |
+++- ESC exits/Alt-U aborts changes ---------------------------+changes -----+

Press a Function key, ENTER to Select Processor, ESC to exit, or Alt-U to abort

offline                 SLC 5/03                           File BEGIN
  PAGE    PAGE
   UP     DOWN
   F1      F2
```

Figure 6-19 Computer screen with 5/03 processor's operating system selections.

9. _____ Press F8, *Exit*.
10. _____ Press F8, *Save and Exit*.
11. _____ You should see the message, "New Archive File Created."
12. _____ Press F9 to save the I/O configuration to file.
13. _____ A screen listing all program files currently on your computer's hard drive. The newly developed program file named *Begin* should be highlighted. Figure 6-20 illustrates a sample screen with the program file *Begin* included.

```
+------------------- SLC-500 ADVANCED PROGRAMMING SOFTWARE -------[ OFFLINE ]-+
|+- PROGRAM DIRECTORY FOR PROCESSOR: SAMPLE    ------------------------------+
||   FILE    PROTECTED       NAME        + C:\IPDS\ARCH\SLC500 ---------------+
||    0                                  |   Name      Size      Date         |
||    1                                  |--------------------------------    |
||    2       Yes                        |   LAB_ONE    929    12-18-97       |
||                                       |   SAMPLE     813    12-18-97       |
||                                       |   BEGIN      813    12-19-97       |
||                                       |                                    |
||                                       |                                    |
||                                       |                                    |
||                                       |                                    |
||                                       |                                    |
||                                       |                                    |
||                                       |                                    |
||                                       |                                    |
||                                       |                                    |
||                                       |                                    |
||                                       |                                    |
++--------------------------------------+- ESC exits/Alt-U aborts changes ----+
NEW ARCHIVE FILE CREATED
Press a Function Key or Enter File Name

offline             SLC 5/03                            File SAMPLE
 OFFLINE                                  CREATE  DEFINE    SAVE
 PRG/DOC                                   FILE    DIR    TO FILE
   F1                                       F6      F7       F9
```

Figure 6-20 Program directory listing the newly created file named *Begin*.

14. _____ If you wanted to develop a new ladder program at this point, you would press F1, *Off-Line Programming / Documentation*. We want to save this processor file to our student floppy disk as our default beginning file for future lab exercises.
15. _____ Press *Escape*.
16. _____ Press F7, *File Options*.
17. _____ Press F7, *Copy to Disk*.
18. _____ Press F5 to select all options.
19. _____ Use the arrow keys to select the *Begin* file to be copied to your student floppy disk. Make sure the *Begin* program is highlighted.
20. _____ Press F3 to select the *Begin* file. Have the floppy in your A drive.
21. _____ Press F1 to begin the operation of copying the *Begin* program from the hard disk to your student floppy.
22. _____ After a few seconds the floppy drive light will go off, and the operation is completed.
23. _____ Press *Escape*. Press *Escape*.
24. _____ Press F3 to return to the main menu.
25. _____ You have now created your default processor file named *Begin* and stored a copy on your student floppy disk. Keep this disk safe as it will be used to create and store lab programming exercises as you continue through this lab manual. Some ladder programs you develop will be saved to your student floppy disk to be edited in later lab exercises.

I/O Configuration Errors and Faulting Your Processor

All APS SLC 500 programming software is very particular about knowing what PLC hardware it is to interact with. If the software detects an invalid address entered while programming, a message will be displayed warning the programmer. If the errors are not corrected, the software will not allow downloading to a PLC processor. If the PLC processor detects a mismatch in the I/O configuration, the processor will fault at the end of the current scan. The following I/O configuration information must match the actual hardware detected by the PLC processor or software as the user ladder program is developed:

1. What processor is being used.
2. For some processors, what operating system is being used.
3. The specific rack or racks that are being used in this PLC configuration.
4. The identity of each I/O module in each and every slot of the PLC rack.

If the programming software detects some portion of the I/O configuration that does not match its configuration, either an error, or an error message, "I/O address not configured," will be displayed when an attempt is made to program an instruction with an unconfigured address. You will not be allowed to download the program into the PLC processor without correcting the errors.

When going into run mode, if the PLC processor detects a hardware and software I/O configuration mismatch, the processor will fault and error codes will be generated.

In the next lab exercise, we will experiment with modifying our I/O configuration so the processor detects an I/O configuration mismatch. We will look at the error codes, and what they mean. In addition we will learn how to correct the problem, clear the faulted processor, and get the PLC back into run mode.

PART II

We will monitor our begin file.

1. _____ Press F3, *Off-Line Programming and Documentation*.
2. _____ F4, *Change File*.

3. _____ Use the arrow keys to select the *Begin* file.
4. _____ F9, *Save to File.*
5. _____ F1, *Off-Line Programming and Documentation.*
6. _____ F2, *Save.*
7. _____ To the message, "File already exists, Overwrite file?" Select F8, *Yes.*
8. _____ F8 to monitor file.
9. _____ Have your personal computer connected to your SLC 500 modular PLC.
10. _____ F1, *Save* and go on line.
11. _____ If you get the message, "Processor name does not match disk file processor name. Continue to download?" press F8, *Yes.*
12. _____ To the message, "Program successfully restored to processor. Change processor to run mode?" press F8, *Yes.*
13. _____ Unplug your PLC.
14. _____ To the message, "Message Timeouts—Loss of communication will be displayed." Two choices will be displayed: F8 to abort program or F10, *No.* DO NOTHING AT THIS TIME.
15. _____ Remove the I/O module in slot three.
16. _____ Restore power to the PLC.
17. _____ Looking at the LEDs on the front of your processor, notice the *Run* light is off and the FAULT (FLT) LED is flashing.
18. _____ Go back to your personal computer and Press F10 to NOT abort program. Pressing F10 should return you to the monitor file screen.
19. _____ Notice the fault code 0352. The fault code is displayed directly above F1, *Change Mode Display* on the terminal screen.
20. _____ Determine what the fault code means by monitoring the processor status file. Press F8, *Data Monitor.*
21. _____ Above the blinking cursor is the message "Enter Data Table Address." Enter the letter *S* for status file.
22. _____ Press *Enter.*
23. _____ Notice the fault code 0352 on the left side of the display.
24. _____ Below the fault code is the fault description, which reads "I/O Module missing from Slot 3." Figure 6-21 (next page) illustrates the status file screen and the fault code being displayed.
25. _____ Correct the problem by removing power from the PLC and reinserting the module.
26. _____ Restore power to the SLC 500.
27. _____ Notice that the fault LED is still flashing, The fault must be cleared.
28. _____ Press F10, *No,* to the message, "Timeouts—loss of communications message. Critical error, enter F8, <yes> to abort program, enter F10 <no> to continue."
29. _____ The fault will be cleared. If the fault does not clear press F10, *Clear Major Fault.*
30. _____ Press *Escape.* This will return you to the file monitor screen.
31. _____ Press F1, *Change Mode.* The processor can now be put into run mode by pressing F3, *Run.*
32. _____ Press F8 to confirm going into run mode.
33. _____ The PLC is put into run mode.

PART III

1. _____ Power down the SLC 500 Modular PLC. Switch the modules in slots three and four. Before the switch, slot three has a 16-point output module, a 1746 OB16. Slot four is a 16-point input module, part number 1746 IB16. After the swap, slot three has the input module and slot four has the output module.

```
ARITHMETIC FLAGS       S:0            Z:0              V:0            C:0

PROCESSOR STATUS       00000000 00000000    SUSPEND CODE                    0
PROCESSOR STATUS       00000000 00000001    SUSPEND FILE                    0
PROCESSOR STATUS       10000000 00000010
                                            WATCHDOG         [x10 ms]:  10
MINOR FAULT            00000000 00000000    LAST SCAN        [x10 ms]:   0
FAULT CODE                         0352     FREE RUNNING CLOCK   00000000 00000000
FAULT DESCRIPTION:I/O module missing from slot 3

MATH REGISTER                  0000 0000

ACTIVE NODE LIST                            I/O SLOT ENABLES
0         10        20        30            0         10        20        30
00000000 00000000 00000000 00000000         11111111 11111111 11111111 11111111

PROCESSOR BAUD RATE            19200        PROCESSOR ADDRESS               1


Press a key or enter value, press Alt-H for help
S:0/0 =
FAULT 0352                no forces           formatted    decimal addr   File SAMPLE
   PAGE     PAGE                         SPECIFY            NEXT   PREV   CLR MIN CLR MAJ
   UP       DOWN        Fault code       ADDRESS            FILE   FILE   FAULT   FAULT
   F1       F2                             F5                F7     F8     F9      F10
```

Figure 6-21 Status screen showing fault code 0352 and explanation.

2. _____ Reapply power.
3. _____ Is the power fault light on?
4. _____ How do you return your programming terminal to be on-line with the PLC?
5. _____ What fault code is displayed?
6. _____ How do we determine what this code signifies?
7. _____ What is the fault description?
8. _____ Power down the PLC and correct problem.
9. _____ Clear fault.
10. _____ Press *Escape* to return to the program edit screen.
11. _____ F3 to exit.
12. _____ F3 to return to main menu.

I/O ERROR PROBLEMS

More information on I/O configuration errors can be found in the SLC 500 Instruction Set Reference Manual. Refer to the SLC status file section, which contains a topic on I/O errors.

Error Codes

If the software detects an incorrect or missing I/O module in the chassis, an error code will displayed. In our case we have the 0355 error code. Error codes are comprised of two parts: The chassis slot and the error code. The rightmost two digits (55) contain the error code. The left two digits represent the chassis slot.

The error code we are looking for is xx55. The x's mean that for determining the error code the x value does not matter. Looking through the list of error codes, find xx55 and read the error explanation. The explanation states that the I/O module has the wrong I/O count. This error can mean two things. First, a slot was configured for, as an example, a 16-point module, but now contains an 8-point module. Also, the processor is looking for 16 outputs but is seeing 16 inputs. No outputs are seen.

Error code	Description
xx52	A module required for the user program is detected as missing or removed.
xx53	When going to run, the user program finds that a slot specified as unused in fact has an I/O module inserted. This code can also mean that an I/O module has reset itself.
	SLC 5/03 only: An attempt was made to enter the run or test modes with an empty chassis.
xx54	A module required for the user program is detected as being of a wrong type.
xx55	A discrete I/O module required for the user program is detected as having the wrong I/O count.
	This code can also mean that a specialty card driver is incorrect.
xx56	The chassis configuration specified in the user program is detected as being incorrect.

Error codes are represented in hexadecimal. Selected I/O error codes are listed in the table above.

I/O Error Slot Determination

The xx characters in the I/O error codes listed above represent the slot number in the PLC chassis where the problem module was found. In our case slot 03 was the first slot found to have an I/O error. To determine which chassis slot contains the module detected in error, substitute for the xx portion of the error code the slot number identified by the hexadecimal code from Figure 6-22.

If the exact slot cannot be determined using a fixed SLC 500 PLC, an error code of 03 will be displayed in the xx position. In using a 5/03 processor, if the exact slot cannot be determined, the xx will be replaced with 1F. Refer to your Instruction Set Reference Manual for additional information on error codes.

	SLOT NUMBERS IN HEXADECIMAL						
SLOT	xx	SLOT	xx	SLOT	xx	SLOT	xx
0	00	8	08	16	10	24	18
1	01	9	09	17	11	25	19
2	02	10	0A	18	12	26	1A
3	03	11	0B	19	13	27	1B
4	04	12	0C	20	14	28	1C
5	05	13	0D	21	15	29	1D
6	06	14	0E	22	16	30	1E
7	07	15	0F	23	17		

Figure 6-22 SLC 500 I/O error slot determination table. Table compiled from Rockwell Automation/Allen-Bradley data.

QUESTIONS

1. Fred, the maintenance person, discovered a bad 1746-A7 chassis on an SLC 500 PLC. The only spare chassis in the plant was a 1746-A10. Fred was careful to get all I/O modules in the same slots. Upon power-up, the processor fault light flashes ON and OFF. Fred comes to you for help. What is the problem? How do you correct this?_____

_____

2. When troubleshooting a problem George discovers an eight-point module in his SLC 500 PLC has failed. The only module available is a similar 16-point module. Upon startup, the processor faults. After attaching to the PLC, you monitor the program file to find a fault code of 0155.

A. Which module is wrong?_____

B. How did you determine this?_____

_____

C. What does the fault code tell you is wrong?_____

_____

D. Since you have no other input modules to install, what do you do now?_____

_____

7

LOADING THE
BEGIN PROGRAM

Each time you start a new ladder program, you will load the default *Begin* file from your student floppy disk to your personal computer's hard drive. When you need to load the *Begin* file from your floppy disk to develop a new program, follow the following steps:

1. _____ Start from the main screen from your APS software.
2. _____ Insert your floppy disk in drive A.
3. _____ Press F7, *File Options*.
4. _____ To copy from floppy select F8, *Copy from Disk*.
5. _____ Select F5, *All of the Above,* to copy entire file to your hard drive.
6. _____ Use the arrow keys to highlight the *Begin* file from the A drive directory.
7. _____ Press F3, *Select File*.
8. _____ Press F1, *Begin Operation*. If you get the message "File already Exists, overwrite file?" F8, *Yes*.
9. _____ The *Begin* file should load from floppy A to your personal computer's hard drive.
10. _____ When the copy procedure is completed, press *Escape* to return to the file operations menu.

RENAME THE *BEGIN* FILE

Rename the default *Begin* file to the file name to be associated with the ladder program you are about to create.

1. _____ Press F3, *Rename*, from the file operations menu.
2. _____ Press F5 to select all options.
3. _____ Select the *Begin* file using the arrow keys.
4. _____ Press F3 to select the source.
5. _____ To enter new file name, press F4, *Enter Destination*.
6. _____ Type in new file name *Ladder_1* at cursor.

7. _____ Press *Enter.*

8. _____ Press F1, *Begin Operation.*

9. _____ The *Begin* file will be renamed. The new file will be highlighted on the drive C directory listing.

10. _____ Press *Escape* to return to the file operations screen.

12. _____ Press *Escape* to return to the main screen.

13. _____ To load the renamed file into the personal computer off-line programming work area so you can edit the file, Press F4, *Change File.*

14. _____ Select file using the up or down arrow keys.

15. _____ Press F9, *Save to File.*

16. _____ The message, "New Configuration saved to file" will be displayed above the blinking cursor on the lower left side of the screen.

17. _____ Press *Escape.*

18. _____ Near the center of the main menu screen you should see current off line file: *Ladder_1.*

19. _____ If you are finished for today press F10 to exit APS software.

20. _____ If you wish to continue, go to the next page, *Exercise 8.*

21. _____ Press F8, *Yes* to accept options.

22. _____ Press F8, *Monitor file* to begin program development.

8

DEVELOPING YOUR FIRST LADDER PROGRAM

This exercise will step you through the procedure to develop your first ladder program using AND logic and the examine-if-closed instruction. Figure 8-1 illustrates the program we will develop.

```
Rung 2:0
    I:1                                                                  O:2
--] [---------------------------------------------------------------( )--
     0                                                                   0

Rung 2:1
    I:1    I:1                                                           O:2
--] [---] [--------------------------------------------------------( )--
     2      3                                                            1

Rung 2:2
    I:1    I:1                                                           O:2
--] [---] [--------------------------------------------------------( )--
     1      4                                                            2

Rung 2:3
    I:1    I:1    I:1                                                    O:2
--] [---] [---] [--------------------------------------------------( )--
     5      6      7                                                     3
```

Figure 8-1   Program for this exercise.

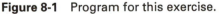

1. _____ Begin program development off-line, by pressing F3, *Off-Line Programming and Documentation.*
2. _____ Press F2, *Save.*
3. _____ F8, *Yes*, to overwrite file. As the program is being saved, you will notice the stars under the column *Size (words)* change to numbers. The process of saving this processor file includes the creation of all default program and data files. This is signified by the stars changing to numerical data identifying the default file size.

4. _____ Press F8, *Monitor File.* A one-rung ladder with the word *End* on the rung will be displayed. The *End* rung is the last rung of the user program. Figure 8-2 illustrates the *End* rung. As a programmer, you will insert rungs into this ladder program. The ladder logic instructions will be placed on each inserted rung.

5. _____ To insert a rung and begin programming, Press F10, *Edit.* The following (Figure 8-2) will be displayed.

```
|                                                             |
|---------------------------------|END+---------------------------------|
|                                                             |
```

```
Press a key for desired editing function
(file     2, rung     0)
offline         no forces              EDITS: NONE              File LADDER_ONE
SAVE/GO  ONLINE   APPEND  INSERT  MODIFY  DELETE   UNDEL  ADVANCD  TEST   CANCEL
ONLINE   CONFIG   RUNG    RUNG    RUNG    RUNG     RUNG   EDITING  EDITS  EDITS
 F1       F2       F3      F4      F5      F6       F7      F8      F9      F10
```

Figure 8-2 The *End* ladder program rung.

6. _____ Press *Insert Rung*, F4. An empty ladder rung will be created above the end rung as illustrated in Figure 8-3.

The function keys on this screen have the following functions:

F1, Branch: This function key takes you to the branch function keys. Branch function keys are used when programming, editing, or deleting branch instructions.

```
I                                                             I
I-------------------------------------------------------------I
I                                                             I
|                                                             |
|---------------------------------|END+-------------------------------|
|                                                             |
```

```
Type a mnemonic or press a key for desired editing function
(file     2, rung     0)
offline         no forces                    RUNG INSERT     File LADDER_ONE
BRANCH            APPEND  INSERT  MODIFY  DELETE   UNDEL          ACCEPT
                  INSTR   INSTR   INSTR   INSTR    INSTR          RUNG
 F1                F3      F4      F5      F6       F7            F10
```

Figure 8-3 Inserted rung on the ladder.

F3, Append Instruction: When programming instructions on a ladder rung, the append instruction function will place the next instruction to the immediate *left* of the current position.

F4, Insert Instruction: When programming instructions on a ladder rung, the insert instruction function will place the next instruction to the immediate *right* of the current position.

F5, Modify Instruction: When you need to modify an instruction already programmed on a ladder rung, the cursor is placed on the instruction to be modified, then the F5 key is pushed.

F6, Delete Instruction: When you need to delete an instruction already programmed on a ladder rung, the cursor is placed on the instruction to be deleted, then the F6 key is pushed.

F7, Undelete Instruction: If you delete an instruction but then change your mind, pressing the F7, *Undelete*, function key will restore the instruction.

F10, Accept Rung: When finished programming, or when editing a ladder rung is completed or accepted by the software, press F10 before going on to the next rung.

7. _____ In order to place an instruction on the rung, Press F4, *Insert Instruction.* The function keys on the display change as shown in Figure 8-4.

```
I                                                                      I
I----------------------------------------------------------------------I
I                                                                      I
!                                                                      !
!----------------------------------------!END+------------------------!
!                                                                      !
```

```
Type a mnemonic or press a key for desired instruction class
(file    2, rung     0)
offline            no forces                    INSTR INSERT   File LADDER_ONE
  BIT    TIMER/    I/O  COMPARE  MATH   MOVE/   FILE  SHIFT/   CONTROL SPECIAL
         COUNTER MESSAGE                LOGICAL       SEQNCER
  F1      F2       F3     F4     F5       F6    F7      F8        F9     F10
```

Figure 8-4   Insert Instruction function key selections.

8. _____ The Insert Instruction screen function keys provide the following instruction options:

F1 will display bit instructions.

F2 will display timer and counter instructions.

F3 will display instructions for sending messages between processors.

F4 will display comparison instructions such as equal to, greater than, or less than instructions.

F5 is where you will find math instructions including add, subtract, multiply, and divide.

F6 is where move and also logical instructions will be found.

F7 includes copy and file fill instructions.

F8 contains shift register and sequencer instructions.

F9 is where you will find subroutine and interrupt instructions.

F10 is for selecting PID and ASCII instructions.

9. _____ If you want to view any of these options, press the specific function key. If you choose to explore any of these other function keys, do not press any of the function keys representing the available instructions. We will be developing ladder programs using most of these function keys in future lab exercises. When you are done looking at other function key selections, press *Escape* to return to the current screen.

10. _____ To begin programming for this exercise, press the F1, *Bit function*, key. The bit instructions will be displayed as illustrated in Figure 8-5.

```
I                                                                        I
I------------------------------------------------------------------------I
I                                                                        I
¦                                                                        ¦
¦------------------------------------------¦END+-------------------------¦
¦                                                                        ¦
```

```
Type a mnemonic or press a key for desired instruction
(file     2, rung      0)
offline            no forces                        INSTR INSERT    File LADDER_ONE
   XIC       XIO       OTE       OTL       OTU       OSR                  OTHERS
  -] [-     -]/[-     -( )-     -(L)-     -(U)-
   F1        F2        F3        F4        F5        F6                    F10
```

Figure 8-5 Insert Instruction function key options.

Referring to Figure 8-1 (page 41) for ladder rung zero: we want to insert a normally open, or XIC, instruction. Press F1, *XIC*. The instruction symbol should display on the ladder rung. The lower left-hand area of your screen will display the following: *Enter Bit address>*.

11. _____ Enter input address I:1/0 for the input instruction. Your screen should look like Figure 8-6 (next page).

12. _____ Press *Enter*.

13. _____ To add the output instruction, press F1, *Bit*.

14. _____ Select the output instruction, press F3, *OTE*.

15. _____ Enter Bit Address> O:2/0. Your screen should look similar to Figure 8-7 (next page).

16. _____ Press *Enter* to accept the address.

17. _____ Press *Escape* to go to the screen with the accept rung function key selection. Your screen should look similar to Figure 8-8 (page 46).

18. _____ Press F10, *Accept Rung*. Add rung one:

19. _____ Press F3 to append a new blank rung below the rung just completed.

20. _____ Press F4 to insert an instruction on the appended rung.

21. _____ Press F1, *Insert Instruction*.

```
I                                                                          I
I--] [----------------------------------------------------------------------I
I                                                                          I
 |                                                                         |
 |------------------------------------|END+----------------------------------|
 |                                                                         |
```

```
Enter the operand
Enter Bit address> I:1/0
```

Figure 8-6   Entering bit address I:1/0.

```
I  I:1                                                                      I
I--] [--------------------------------------------------------------( )--I
I   0                                                                       I
 |                                                                         |
 |-------------------------------|END+-------------------------------------|
 |                                                                         |
```

```
Enter the operand
Enter Bit address> O:2/0
offline          no forces                    INSTR INSERT    File Ladder_One
```

Figure 8-7   Entering output bit address O:3/1.

22. _____ Press F1, *XIC*. This will add the examine if closed instruction symbol.
23. _____ Enter Bit Address> *I:1/2* and press *Enter* to accept the address. Add an instruction in series with the I:2/2 XIC instruction.
24. _____ Press F1, *Bit*.
25. _____ Press F1, *XIC*. This instruction is placed in series with the previous instruction.
26. _____ Enter Bit Address>*I:1/3*. Press *Enter* to accept address. Add the output instruction.
27. _____ Press F1, *Bit*.
28. _____ Press F3, *OTE*. This is the output enable instruction.
29. _____ Enter Bit Address>*O:2/1*.
30. _____ Press *Enter* to accept the output address.

```
I   I:1                                                                        O:2   I
I--] [------------------------------------------------------------------( )--I
I     0                                                                          O   I
:                                                                                    :
:-----------------------------------------!END+--------------------------------------:
:                                                                                    :
```

```
Type a mnemonic or press a key for desired editing function
(file      2, rung      0)
offline           no forces                              RUNG INSERT   File Ladder_One
BRANCH            APPEND  INSERT  MODIFY  DELETE  UNDEL                        ACCEPT
                  INSTR   INSTR   INSTR   INSTR   INSTR                         RUNG
   F1              F3      F4      F5      F6      F7                           F10
```

Figure 8-8 Accepting the completed rung.

31. _____ Press *Escape.*
32. _____ Press F10 to accept the completed rung. Notice the software automatically appended a new rung for you to program.
33. _____ Press F4, *Insert Instruction,* to insert the first instruction on this new rung.
34. _____ Press F1, *Bit.*
35. _____ Press *XIC.*
36. _____ Enter Bit Address>*I:1/1.*
37. _____ Press *Enter* to accept the address.
38. _____ To add the next instruction in series with the first, press F1, *Bit.*
39. _____ Press F1, *XIC.*
40. _____ Enter Bit Address>*I:1/4.*
41. _____ Press *Enter* to accept the address.
42. _____ To add the output instruction, press F1, *Bit.*
43. _____ Press F3, *OTE.*
44. _____ Enter Bit Address>*O:2/2.*
45. _____ Press *Enter* to accept the address.
46. _____ Press *Escape.*
47. _____ Press F10 to accept the rung.

Adding instructions to rung three:

48. _____ Press F4, *Insert Instruction.*
49. _____ Press F1, *Bit.*
50. _____ Press F1, *XIC.*
51. _____ Enter Bit Address>*I:1/5.*
52. _____ Press *Enter* to accept the address.
53. _____ To enter the next instruction in series with the first, press F1 to select a bit instruction.
54. _____ Press F1, *XIC.*

55. _____ Enter Bit Address>*I:1/6*.
56. _____ Press *Enter* to accept the address.
57. _____ Enter the next instruction in series with the previous. Press F1 to select a bit instruction.
58. _____ Press F1, *XIC*.
59. _____ Enter Bit Address>*I:1/7*.
60. _____ Press *Enter* to accept the address.
61. _____ Enter the output instruction. Press F1, *Bit*.
62. _____ Press F3, *OTE*.
63. _____ Enter Bit Address>*O:2/3*.
64. _____ Press *Enter* to accept the address.
65. _____ Press *Escape*.
66. _____ Press F10 to accept the rung.
67. _____ This completes this ladder program. Since the next rung was automatically added for you, press *Escape* to delete this extra rung.
68. _____ Connect your personal computer to your SLC 500 or MicroLogix PLC as directed by your instructor.
69. _____ Press F1, *Save / Go On-line*.
70. _____ Answer *Yes*, by pressing F8, to the message: "File already exists, overwrite file?" Accept the default selections displayed in the menu box that popped up.
71. _____ If you get the message: "Processor program name does not match file program name. Continue with download?" Press F8, *Yes*. This message alerts you that the program you just developed on your personal computer, which is stored on the hard drive, does not match the program in the PLC processor. Your answering *yes* causes the program you just completed to be downloaded and overwrite the current PLC program. Being overwritten, the old PLC program is lost forever.
72. _____ If you get the message: "Attempting to restore file to processor, processor must be in program mode. Change processor mode to program?" press F8, *yes*. With the processor in program mode, the download process can continue.
73. _____ When you get the message, "Program successfully restored to processor. Change processor to Run?" Press F8, *Yes*. The processor will now go into run mode.
74. _____ Test your inputs to verify they correctly control their respective outputs.
75. _____ When completed testing your program, Press F3, *Exit*, to return to program directory for processor: *Ladder_1*.
76. _____ Press F3, *Return to main menu* and go off-line.
77. _____ Press F7, *File operations*. You will now save your program named *Ladder_1* to your student floppy disk.
78. _____ Press F7, *Copy to Disk*.
79. _____ Select all of the options by pressing F5.
80. _____ Using arrow keys, select your processor file *Ladder_1*.
81. _____ Press F3 to select the file.
82. _____ Press F1 to begin the operation of copying your processor file from the personal computer hard drive to your student floppy disk.
83. _____ When copying is completed, press *Escape*.
84. _____ Press *Escape* to return to the program directory for processor screen.
85. _____ Press F3 to return to the main menu.
86. _____ If you are finished programming for today, press F10 to exit the software and return to your computer's operating system.

9

PRACTICE WITH YOUR FIRST LADDER PROGRAM

For this exercise you will develop your first ladder program. Figure 9-1 illustrates the program you will develop.

```
Rung 2:0
|   I:1     I:1                                                          O:2 |
|--] [---] [-----------------------------------------------------------( )--|
|     1       3                                                          2  |

Rung 2:1
|   I:1                                                                  O:2 |
|--] [-----------------------------------------------------------------( )--|
|     5                                                                  1  |

Rung 2:2
|   I:1     I:1     I:1                                                  O:2 |
|--]/[---]/[---]/[-----------------------------------------------------( )--|
|     0       2       4                                                  0  |

Rung 2:3
|   I:1     I:1     I:1     I:1                                          O:2 |
|--]/[---] [---] [---] [-----------------------------------------------( )--|
|     5       6       3       1                                          3  |

Rung 2:4
|   I:1                                                                  O:2 |
|--] [-----------------------------------------------------------------( )--|
|     2                                                                  4  |

Rung 2:5
|   I:1     I:1                                                          O:2 |
|--] [---] [-----------------------------------------------------------( )--|
|     7       2                                                          5  |

Rung 2:6
```

Figure 9-1   Your first ladder program.

1. _____ Copy the *Begin* file from your student floppy disk to your personal computer's hard drive.
2. _____ Rename the *Begin* file to *Ladder_2*.
3. _____ Escape to the main program screen. Press F3 *Off line programming/documentation*.
4. _____ Select file by pressing F4, *Change File*.
5. _____ Use arrow keys to select *Ladder_2* file.
6. _____ Press F9, *Save To File*.
7. _____ F1, *Off line programming/ documentation*.
8. _____ Press F8, *Monitor File*. A one-rung ladder with the word *End* on the rung will be displayed.
9. _____ F10 *Edit*.
10. _____ Using your newly learned programming skills, complete the program illustrated in Figure 9-1. Make sure you adjust the I/O addresses, if necessary, so they are correct for your particular PLC.
11. _____ When completed programming this ladder program.
12. _____ F1, *Save and go online*.
13. _____ F8, *Yes* to the message, "File already exists. Overwrite file?"
14. _____ If you get the message, "Processor Program Name does not match disk File Program Name. Continue with download?" F8, *Yes*.
15. _____ Change processor to Run Mode, F8, *Yes*.
16. _____ Test to verify that inputs control outputs as expected.
17. _____ When finished, F3, *Exit*.
18. _____ F7, *File options*.
19. _____ F7, *Copy to disk*.
20. _____ F5, *All of the above*.
21. _____ Select *Ladder_2*.
22. _____ F3, *Select file*.
23. _____ F1, *Begin operation*.
24. _____ *Escape*.
25. _____ *Escape*.
26. _____ F3, *Return to menu*.
27. _____ F10 to exit system, or go on to Lab 10.

10

DEVELOPING A PARALLEL LOGIC LADDER PROGRAM

This exercise will step you through the procedure to develop a ladder program with parallel logic. Figure 10-1 illustrates the program we will develop.

1. _____ Copy the *Begin* file from your student floppy disk to your personal computer's hard drive.
2. _____ Rename the *Begin* file to *Ladder_3*.

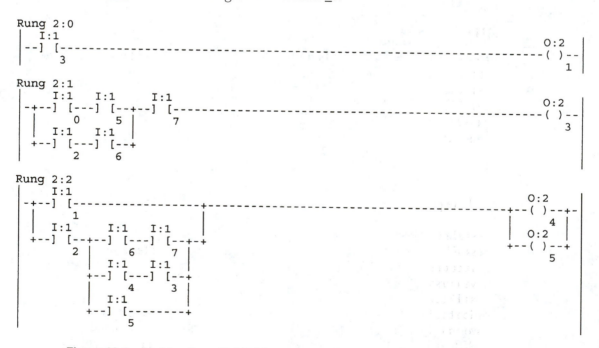

```
Rung 2:0
    I:1                                                           O:2
--] [------------------------------------------------------------( )--
    3                                                             1

Rung 2:1
    I:1    I:1     I:1                                            O:2
-+--] [---] [--+--] [-------------------------------------------( )--
 |   0     5  |   7                                               3
 |  I:1    I:1|
 +--] [---] [--+
     2     6

Rung 2:2
    I:1                                                           O:2
-+--] [--------------------+-----------------------------------+--( )--+-
 |   1                     |                                   |        4
 |  I:1     I:1    I:1     |                                   |  O:2
 +--] [--+--] [---] [--+--+                                   +--( )--+
     2   |   6     7  |                                              5
         |  I:1    I:1|
         +--] [---] [--+
             4     3
         |  I:1       |
         +--] [--------+
             5
```

Figure 10-1 *Ladder_3* parallel ladder program.

3. _____ Escape to the main screen.
4. _____ F3, *Offline program / documentation*.
5. _____ Select file by pressing F4, *Change File*.
6. _____ Use arrow keys to select *Ladder_3* file.
7. _____ Press F9, *Save To File*.
8. _____ F1, *Offline program / documentation*.
9. _____ Press F8, *Monitor File*. A one-rung ladder with the word *End* on the rung will be displayed. The end rung is the last rung of the user program. As a programmer, you will insert rungs into this ladder program. The ladder logic instructions will be placed on each inserted rung.

PROGRAMMING RUNG ZERO

10. _____ To insert a rung and begin programming, Press F10, *Edit*.
11. _____ Press *Insert Rung*, F4. An empty ladder rung will be created above the end rung.
12. _____ Press F4, *Insert Instruction*.
13. _____ Press F1, *Bit*.
14. _____ Press F1, *XIC*.
15. _____ Enter *bit address> I:1/3*.
16. _____ Press *Enter* to accept the address.
17. _____ Press F1, *Bit*.
18. _____ Press F3, to put the output instruction, *OTE*, on the rung.
19. _____ Enter a bit address> O:2/1.
20. _____ Press *Enter* to accept the address.
21. _____ Press *Escape*.
22. _____ Press F10 to accept the rung.

PROGRAMMING RUNG ONE

1. _____ Press F3, *Append Rung*. This will add the new rung you will add instructions to.
2. _____ Press F4, *Insert Instruction*.
3. _____ Press F1, *Bit*.
4. _____ Press F1, *XIC*.
5. _____ Enter *bit address> I:1/0*.
6. _____ Press *Enter* to accept the address.

Add a Parallel Branch

Next we will add the parallel branch as illustrated in Figure 10-1, rung one.

1. _____ Press *Escape* to display the branch function.
2. _____ Press F1, *Branch,* to start the programming sequence to enter a branch, or parallel instructions, to your ladder rung.
3. _____ Move cursor with left arrow key to left power rail.
4. _____ Press F4, *Insert Branch*.
5. _____ Notice the letter *A* appears on the left power rail, and the letter *B* appears on the right power rail. Since our cursor is already in the *A* position, and we want to branch around the XIC, I:1/0 instruction, select target *B*. The branch will start at target *A* and go to target *B*.

6. _____ Press F2, *Select Target B*.
7. _____ The cursor should be on the right-side connection of the branch back to the main rung. We need to program an XIC instruction on the parallel branch to the left of the cursor.
8. _____ Press F4, *Insert Instruction*.
9. _____ Press F1, *Bit*.
10. _____ Press F1, *XIC*.
11. _____ Enter bit address> *I:1/2*.
12. _____ Press *Enter* to accept the address.
13. _____ Next we will add an instruction in series with the first XIC instruction on the parallel branch.
14. _____ Press F1, *Bit*.
15. _____ Press F1, *XIC*.
16. _____ Enter a bit address> *I:1/6*.
17. _____ Use the up arrow key to move the cursor to the intersection of the right vertical connection from the parallel branch and the main rung.
18. _____ Refer to Figure 10-1 (page 50). To program the XIC instruction I:1/5, press F1, *Bit*.
19. _____ Press F1, *XIC*.
20. _____ Enter bit address> *I:1/5*.
21. _____ Press *Enter* to accept the address.
22. _____ Next enter the *XIC, I:1/4* instruction.
23. _____ Press F1, *Bit*.
24. _____ Press F1, *XIC*.
25. _____ Referencing Figure 10-1, notice the instruction was placed on the wrong side of the branch. We need to delete this incorrect instruction, and program the XIC, I:1/7 instruction on the right side of the branch.
26. _____ Press *Escape* to remove the incorrect instruction.
27. _____ Press arrow key to move cursor out of parallel branch to the right power rail.
28. _____ Press F1, *Bit*.
29. _____ Press F1, *XIC*.
30. _____ Enter bit address> *I:1/7*.
31. _____ Press *Enter* to accept the address.
32. _____ Press F1, *Bit*.
33. _____ Press F3, *OTE*.
34. _____ Enter bit address> *O:2/3*.
35. _____ Press *Escape*.
36. _____ Press F10, *Accept rung*.

RUNG TWO

1. _____ If the programming software does not automatically append a new rung, press F3, *Append Rung*. This will add the new rung you will add instructions to.
2. _____ Press F4, *Insert Instruction*.
3. _____ Press F1, *Bit*.
4. _____ Press F1, *XIC*.
5. _____ Enter a bit address> *I:1/1*.
6. _____ Press *Enter* to accept the address.

Add a Parallel Branch

Next we will add the parallel branch as illustrated in Figure 10-1.

1. _____ Press escape to display the branch function.
2. _____ Press F1, *Branch,* to start the programming sequence to enter a branch, or parallel instructions to *I:1/1* on your ladder rung.
3. _____ Cursor should be on the right power rail.
4. _____ Press F4, *Insert Branch.*
5. _____ Notice the letter *A* appears on the left power rail, and the letter *B* appears on the right power rail. Since our cursor is already in the *B* position, and we want to branch around the *XIC, I:1/1* instruction, select target *A*. The branch will start at target *A* and go to target *B*.
6. _____ Press F1, *Select Target A.*
7. _____ The cursor should be on the right-side connection of the branch back to the main rung. We need to program an XIC instruction on the parallel branch to the left of the cursor.
8. _____ Press F4, *Insert Instruction.*
9. _____ Press F1, *Bit.*
10. _____ Press F1, *XIC.*
11. _____ Enter bit address> *I:1/2.* Notice this instruction has already been used on rung one. There is no reason why an input instruction cannot be used more than once in a ladder program.
12. _____ Press *Enter* to accept the address.
13. _____ Next we will add an instruction in series with the first XIC instruction on the parallel branch.
14. _____ Press F1, *Bit.*
15. _____ Press F1, *XIC.*
16. _____ Enter a bit address> *I:1/6.* Notice this instruction has already been used on rung one. There is no reason why an input instruction cannot be used more than once in a ladder program.
17. _____ Press *Enter* to accept the address.
18. _____ Press F1, *Bit.*
19. _____ Press F1, *XIC.*
20. _____ Enter bit address *I:1/7.*
21. _____ Press *Enter* to accept the address.
22. _____ Press *Escape* to return to the screen with the branch function.
23. _____ Press F1, *Branch.*
24. _____ Press F4, *Insert Branch.* Notice the letters *A, B, C,* and *D* appear on the portion of their ladder branch we are currently working on. Each letter is called a target for the branch. Target A is at the left vertical connection of the branch. Target B is between instructions addressed as *I:1/2* and *I:1/6.* While target *C* is between instructions addressed as *I:1/6* and *I:1/7.* Target D is the right vertical power connection of the branch.
25. _____ Referring to Figure 10-1 (page 50), our branch needs to connect between instruction address *I:1/2* and *I:1/6.* This is target *B.* Press F2, *Target B,* to complete the branch connection.
26. _____ Press F4, *Insert instruction.*
27. _____ Press F1 to select a bit input instruction.
28. _____ The program example illustrates an XIC instruction, Press F1, *XIC.*
29. _____ Enter a bit address> *I:1/4.*
30. _____ Press *Enter* to accept the address.
31. _____ We need to enter an XIC instruction in series with *I:1/4.* Press F1, *Bit.*
32. _____ To place the XIC instruction on the branch press F1, *XIC.*
33. _____ Enter a bit address> *I:1/3.*

EXPERIMENTATION WITH BRANCHING

The next few programming activities will give you practice entering different parallel branches and then deleting them. This will give you practice in deleting mistakes that might be made while programming branches.

1. _____ Cursor back to the last XIC instruction that you just added, *I:1/3*.
2. _____ Press *Escape*.
3. _____ Let's try adding another branch to this rung.
4. _____ Press F1, *Branch*.
5. _____ To insert the new branch, press F4, *Insert Branch*.
6. _____ Press F1 to select Target *A*. Notice the branch created. Let's assume that we do not wish to keep this last branch. We need to delete the incorrect branch.
7. _____ Press F1, *Branch*. This will take you to the branch function keys.
8. _____ Press F6, to delete the unwanted branch.
9. _____ The last branch entered should be deleted.
10. _____ Let's try adding another branch around the same XIC contact. Press F1, *Branch*.
11. _____ F4, *Insert Branch*.
12. _____ This time select F2, *Target B*.
13. _____ Let's assume that we do not want this branch either.
14. _____ Press F1, *Branch*. This will take you to the branch function keys.
15. _____ Press F6, *Delete Branch*, to delete the unwanted branch.
16. _____ The last branch entered should be deleted.
17. _____ Let's try one more branching option. With the cursor on I:1/3 instruction, press F1, *Branch*.
18. _____ Press F4, *Insert Branch*.
19. _____ Press F3, *Target C*. You can see how this selection works.
20. _____ Let's remove this branch too, remember how?
21. _____ Now that you are back to the original rung before we started experimenting with the branch options available, let's add the branch that we really wanted. Press F1, *Branch*.
22. _____ Press F2, *Extend down*. We want to branch around the I:1/4 and I:1/3 instructions.
23. _____ Press F3, *Append instruction on new branch*.
24. _____ Press F1, *Bit*.
25. _____ Press F1, *XIC*.
26. _____ Enter a bit address> *I:1/5*.
27. _____ Next we will enter parallel output instructions (OTE).
28. _____ Cursor to the right power rail.
29. _____ To select the OTE Instruction, Press F1, *Bit*.
30. _____ Press F3, *OTE* (output instruction).
31. _____ Enter a bit address> *O:2/4*.
32. _____ Press *Enter* to accept the address.
33. _____ Move cursor to the right power rail.
34. _____ To return to the branch selection screen, press *Escape*.
35. _____ Press F1, *Branch*.
36. _____ Press F4, *Insert Branch*. Notice that target *A* is the left power rail. Target *B* is the left side of the previous OTE instruction, while target *C* is the right power rail. Since we need to program a parallel output, select target *B*.
37. _____ Press F2, Target *B*.
38. _____ Press F4, *Insert Instruction*.
39. _____ Press F1, *Bit*.
40. _____ Press F3, to program the OTE instruction on the parallel branch.

41. _____ Enter a bit address> *O:2/5*.
42. _____ To accept the completed rung press *Escape*.
43. _____ Press F10, to accept and complete your rung.
44. _____ Press *Escape* to remove the automatically appended blank rung.
45. _____ Hook up your personal computer to your PLC as directed by your instructor.
46. _____ Press F1, *Save and go on-line*.
47. _____ Put the PLC processor into run mode.
48. _____ Test your program operation.
49. _____ Don't forget to save this program as *Ladder_3* on your student floppy disk.

11

PRACTICE WITH THE PARALLEL LOGIC LADDER PROGRAM

For this exercise you will develop your own first ladder program with parallel logic. Figure 11-1 on the next page illustrates the program you will develop.

1. _____ Copy the *Begin* file from your student floppy disk to your personal computer's hard drive.
2. _____ Rename the *Begin* file to *Ladder_4*.
3. _____ *Escape*.
4. _____ Press F3, *Offline programming / documentation*.
5. _____ Select file by pressing F4, *Change File*.
6. _____ Use arrow keys to select *Ladder_4 file*.
7. _____ Press F9, *Save To File*.
8. _____ Press F1, *Offline programming / Documentation*.
9. _____ Press F8, *Monitor File*. A one-rung ladder with the word *End* on the rung will be displayed.
10. _____ Using your newly learned programming skills, complete the program as illustrated in Figure 11-1. Make sure you adjust the I/O addresses, if necessary, so they are correct for your particular PLC.
11. _____ When completed programming this ladder program, download and run the program to verify proper operation.
12. _____ When completed testing your program, save it to your student floppy.

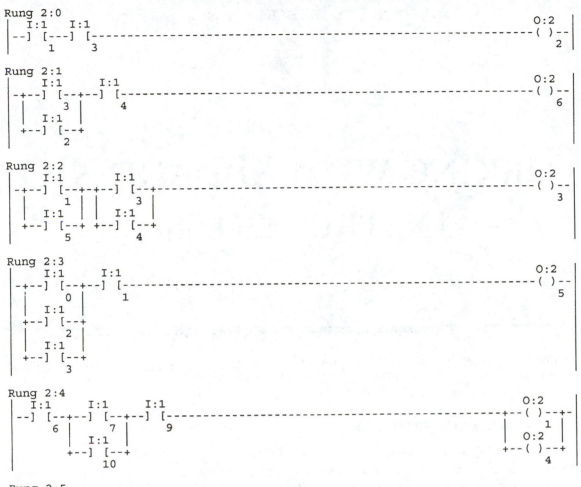

Figure 11-1 *Ladder_4* parallel logic ladder program.

12

WORKING WITH XIC AND XIO INSTRUCTIONS

For this exercise you will experiment with two ladder rungs. One rung will have an XIC instruction and the other rung an XIO instruction. We will learn about how these instructions react to input conditions through a couple of experiments.

ADDITIONAL MATERIALS NEEDED

You will need either a MicroLogix 1000, fixed SLC 500, or modular SLC 500 PLC with a toggle switch connected to input I:1/0, and a toggle switch connected to input I:1/1. If you are using a MicroLogix, or fixed PLC, connect one toggle switch to inputs I:0/0 and I:0/1.

THE LAB

With your toggle switches connected to the two inputs, develop the following program. When completed, save and go on-line, and put the processor in run mode.

PART I

1. _____ Copy the *Begin* file from your student floppy disk to your personal computer's hard drive.

```
Rung 2:0
|   I:1                                                                    O:2  |
|--] [---------------------------------------------------------------------( )--|
|    0                                                                       0  |
Rung 2:1
|   I:1                                                                    O:2  |
|--]/[---------------------------------------------------------------------( )--|
|    1                                                                       1  |
```

Figure 12-1 Working with XIC and XIO instructions, *Ladder_5* program.

2. _____ Rename the *Begin* file *Ladder_5*.
3. _____ Press the *Escape* key twice.
4. _____ Select file by pressing F4, *Change File*.
5. _____ Use arrow keys to select *Ladder_5* file.
6. _____ Press F9, *Save To File*.
7. _____ Press F1, *Offline programming and documentation*.
8. _____ Press F8, *Monitor File*. A one rung ladder with the word *End* on the rung will be displayed. The end rung is the last rung of the user program.
9. _____ To insert a rung and begin programming, Press F10, *Edit*.

Programming Rung Zero

10. _____ Press *Insert Rung*, F4. An empty ladder rung will be created above the end rung .
11. _____ Press F4, *Insert Instruction*.
12. _____ Press F1, *Bit*.
13. _____ Press F1, *XIC*.
14. _____ Enter a bit address> *I:1/0*.
15. _____ Press *Enter* to accept the address.
16. _____ Press F1, *Bit*.
17. _____ Press F3, to put the output instruction, OTE, on the rung.
18. _____ Enter a bit address> *O:2/0*.
19. _____ Press *Enter* to accept the address.
20. _____ Press *Escape*.
21. _____ Press F10 to accept the rung.

Programming Rung One

1. _____ Press F3, *Append Rung*. This will add the new rung you will add instructions to.
2. _____ Press F4, *Insert Instruction*.
3. _____ Press F1, *Bit*.
4. _____ Press F1, *XIO*.
5. _____ Enter a bit address> *I:1/1*.
6. _____ Press *Enter* to accept the address.
7. _____ Press F1, *Bit*.
8. _____ Press F3, *OTE* to add the output instruction.
9. _____ Enter bit address> *O:2/1*.
10. _____ Press *Enter* to accept the address.
11. _____ You can press either *Escape* or *Enter* to go to the *Accept the rung* function.
12. _____ Press F10 to accept the rung.
13. _____ Press *Escape* to remove the automatically generated empty rung.
14. _____ Connect your PLC to your personal computer as directed by your instructor.
15. _____ Press F1, *Save/Go On Line*.
16. _____ Answer *Yes*, F8, to the message, "File already exists. Overwrite File?"
17. _____ If you get the message: "Processor program does not match disk file program mode. Continue with download?" Answer *Yes* by pressing F8.
18. _____ Press F8 to the "Change processor mode to Run?" question.

Analysis of Ladder Program Behavior

1. _____ Verify that both toggle switches are open.
2. _____ The XIC instruction on rung zero is false. Output O:2/1 is off.
3. _____ The XIO (examine if open) instruction on rung one is true and output O:2/1 is ON.

The XIO is a normally closed instruction. If you think about a mechanical relay, the normally closed contacts pass power whenever the switching device controlling the relay's coil is not closed. This same principle holds true for this situation, too. Here the input toggle switch connected to I:1/1 is not closed, and the normally closed instruction is true.

4. _____ Let us look at the input status data file, file one. We can observe the ON or OFF status of our two inputs by looking at their status bits.

5. _____ The easiest way to view the data table for one of the input instructions is use the arrow keys to move the cursor on to one of the input instructions.

6. _____ Press F8, *Data Monitor.*

7. _____ The input status data file should display.

8. _____ Your screen should display data similar to that illustrated below. As you view this, keep in mind this input status display is only valid for a modular SLC 500 PLC configured like the default SLC 500 PLC used to develop the labs. Your screen may be slightly different.

```
Address:        15    Data    0
I:1          00000000 00000000
I:4                   00000000
```

This status table display tells us that there is a 16-point input module in slot one, address I:1. There is one zero signifying each input point. An eight-point module is in slot four, address I:4. There is one zero signifying each of the eight input points.

The cursor should be on the bit address corresponding to the instruction on which you placed the cursor in the monitor file screen. All bits should be zeros as all inputs are OFF.

9. _____ Now let us look at the output status data file, file zero. To view data file zero, press F8, *Previous file.*

10. _____ The data file should look similar to the data below. Again remember this data display is I/O configuration dependent. This I/O configuration reflects the default PLC used to develop these lab exercises.

```
Address:        15    Data    0
O:2          00000000 00000010
O:3          00000000 00000000
O:5          00000000 00000000
O:6                   00000000
```

This data display tells you there is:

A 16-point output module in slot two, address O:2
A 16-point output module in slot two, address O:3
A 16-point output module in slot two, address O:5
A 8-point output module in slot two, address O:6

11. _____ Note that output O:2, bit 1 (second bit from the right), is a 1. This signifies that output address O:2/1 is true. The LED marked 1, on the output module in slot two, should also be ON. The ON LED is telling the module to turn power ON to the output screw terminal on the output module addressed as O:2/1.

12. _____ To specify a specific data table address you may want to see, press the F5 key, *Specify Address.* As an example, to view address I:1/1 (the input data file), press F5.

13. _____ The cursor should be blinking below "Enter data table address:"

14. _____ Enter the address *I:1/1*.

15. _____ Press *Enter.*

16. _____ The screen should change to display the input data screen. The cursor should be on address I:1/1.

17. _____ Press F7, *Next file,* to view the first page of the status file.

18. _____ Press F7 to view the *Bit file, File three.* You should see no data in binary or bit file three.

19. _____ Press F7 to view *Timer file, File four.* There should be no data in timer file four.

20. _____ Press F7 to view *Counter file, File five.* There should be no data in file five.

21. _____ Press F7 to view *Control file, File six.* There should be no data in file six.

22. _____ Press F7 to view *Integer file, File seven.* There should be no data in the integer file.

23. _____ Press F7 to view *Float (floating point) file, File eight.* There should be no data in the floating point file.

24. _____ Pressing F7 again will result in the message "No more data." Only the default data files, zero through eight, have been created. We will create additional data files in later labs.

25. _____ You can press the F8 key to step back through the individual files.

26. _____ Pressing F5 will prompt you to enter the address of a specific data file address you wish to view.

27. _____ Press *Escape* when finished looking at the data files, to return to editing your ladder program.

28. _____ With both toggle switches open fill in the table in Figure 12-2:

Toggle Switch	Input	Instruction	Output ON or OFF?
Open	I:1/0	XIC	
Open	I:1/1	XIO	

Figure 12-2   XIC and XIO instruction behavior with open field devices.

29. _____ With both toggle switches still open fill in the table in Figure 12-3.

30. _____ Close the toggle switch connected to input I:1/0. Does I:1, bit 0, turn to a 1?

Toggle switch	Input	Instruction	Input Status File Bit Condition	Output Status File Bit Condition	Output ON or OFF ?
Open	I:1/0	XIC			
Open	I:1/1	XIO			

Figure 12-3   XIC and XIO instruction interaction with I/O status files.

31. _____ Press F8, *Previous file.* Has the output bit, O:2 bit 0, changed to a 1?

32. _____ The output LED marked 0 on the output module, or output section of your MicroLogix 1000 or fixed SLC 500 PLC, should be ON.

33. _____ Press F7, *Next File* to return to the input data file.

34. _____ Energize the input toggle switch connected to I:1/1. This is the input referenced to the XIO (examine if open, or normally closed) instruction on rung one.

35. _____ The input status data file had a 0 in address I:1/1 when the toggle switch was sending no input signal into the input module. Is the data table input status bit now a 0?

36. _____ Energizing the toggle switch connected to input address I:1/1, you should see the following happen:

1. LED #1 on the input module should come ON. This signifies that the input module has seen an input signal.
2. The input status bit in the input status data table should change from a 0 to a 1.
3. The XIO input instruction on the ladder rung should become false. Being false, the instruction will not be intensified any longer.
4. Output data file address O:2/1 should change from a 1 to a 0.
5. LED #1 on the output module, or output section should turn off.

37. _____ With both toggle switches closed fill in the table in Figure 12-4.

Toggle Switch	Input	Instruction	Output ON or OFF?
Closed	I:1/0	XIC	
Closed	I:1/1	XIO	

Figure 12-4   XIC and XIO instruction behavior with closed field devices.

38. _____ With both toggle switches still closed, fill in the table in Figure 12-5.

Toggle Switch	Input	Instruction	Input Status File Bit Condition	Output Status File Bit Condition	Output ON or OFF ?
Closed	I:1/0	XIC			
Closed	I:1/1	XIO			

Figure 12-5   XIC and XIO instruction interaction with I/O status files.

PART II

Part two of this exercise will edit your current program. The edited program is illustrated in Figure 12-6.

```
Rung 2:0
   I:1                                                                      O:2
--] [--------------------------------------------------------------------( )--
    0                                                                       0

Rung 2:1
   I:1                                                                      O:2
--]/[--------------------------------------------------------------------( )--
    0                                                                       1

Rung 2:2
   I:1                                                                      O:2
--]/[--------------------------------------------------------------------( )--
    1                                                                       3

Rung 2:3
   I:1                                                                      O:2
--] [--------------------------------------------------------------------( )--
    1                                                                       2
```

Figure 12-6   Edited program for lab exercises.

To go off-line and edit our ladder program, complete the following steps:

1. _____ Press F10, *Edit*.
2. _____ You will be asked: "Which program do you want to edit?" You have the following function key selections: F1, *Off-Line Processor;* F3, *Off-Line Disk;* and F7, *Online Edit*. Each function key choice is defined below:

F1, Off-Line Processor: Editing the PLC processor file off-line entails uploading the current processor file from the PLC and storing the file on your personal computer hard drive. This option reads the current PLC program, while the PLC processor is in run mode, and uploads the processor file to the hard drive. This option provides current data table values at the time of upload.

F3, Off-Line Disk: Off-line disk selection takes the personal computer off line with the PLC's processor and edits the program stored on the personal computer's hard drive. This option does not provide current data table data.

F7, Online Edit: On-line editing is available only if you have a 5/03 or 5/04 PLC processor. On-line editing allows the programmer to make limited programming changes to the PLC processor ladder program while the processor is in run mode and the programmer is on-line monitoring the running PLC program.

3. _____ For this exercise, select F3 to edit the off-line personal computer's hard disk program. Notice the PLC remains in run mode. You are now off-line and not communicating with the PLC.

4. _____ Use the arrow keys to highlight rung one.

5. _____ Press F3, *Append Rung.*

6. _____ Press F4, *Insert Instruction.*

7. _____ Press F1, *Bit.*

8. _____ Press F1, *XIC.*

9. _____ Enter Bit Address > *I:1/1.*

10. _____ Press *Enter* to accept the address.

11. _____ Press F1, *Bit.*

12. _____ Press F3, *OTE.*

13. _____ Enter Bit Address > *O:2/2.*

14. _____ Press *Enter* to accept the address.

15. _____ Press *Escape* or *Enter* to go to the *Accept Rung* function.

16. _____ Press F10, *Accept Rung.*

17. _____ Press *Escape* if a rung is automatically appended.

18. _____ Move the cursor so it is on rung one.

19. _____ Press F4 to insert the next rung. Notice that *inserting* a rung places it ahead of the selected rung, while *appending* places the new rung after the current rung.

Enter Rung Three:

20. _____ Press F4, *Insert Instruction.*

21. _____ Press F1, *Bit.*

22. _____ Press F2, *XIO.*

23. _____ Enter Bit Address > *I:1/0.*

24. _____ Press *Enter* to accept the address.

25. _____ Press F1, *Bit.*

26. _____ Press F3, *OTE.*

27. _____ Enter Bit Address > *O:2/1.*

28. _____ Press *Escape* or *Enter* to go to the *Accept Rung* function.

29. _____ Press F10, *Accept Rung.*

30. _____ Press *Escape* if the next rung is automatically appended

31. _____ Notice that rungs one and two both have the same output address: O:2/1. We need to change one of these output addresses.

32. _____ Move the cursor to rung two, OTE, O:2/1.

33. _____ F5, *Modify rung.*

34. _____ F5, *Modify instruction.* Notice that the original rung is identified with R's on the power rails, while the replacement rung is identified with I's.

35. _____ Notice the cursor blinking next to the instruction type, *OTE*.
36. _____ Since we do not wish to modify the instruction type, press *Enter*.
37. _____ The instruction address should be displayed.
38. _____ Backspace to deleted the 1.
39. _____ Enter the number 3. This should change the address to O:2/3.
40. _____ Press *Enter*.
41. _____ Press *Escape*.
42. _____ F10 to accept the rung.
43. _____ We will now download our edited ladder program and experiment with input behavior in relation to switch position.
44. _____ Press F1, *Save and Go On Line*.
45. _____ To the message: "File already exists; overwrite file?" Press F8, *Yes*. This simply writes the edited ladder program from the computer hard drive over the old program residing in PLC processor memory.
46. _____ If your processor is in run mode, you will get the message: "Attempting to restore file to processor, processor must be in program mode. Change processor to program mode?" Answer F8, *Yes*.
47. _____ Change processor to Run mode.
48. _____ Close input switch 0.
49. _____ Leave input switch 1 open.
50. _____ As the program runs, experiment with the normally open push button to see how its physical position associates with the ladder program's instructions.
51. _____ Fill in the table in Figure 12-7 as you observe how the input instructions react to the toggle switches' open or closed conditions.

NORMALLY OPEN PUSH BUTTON ADDRESS I:1/0			
Push Button	—I I— True or False?	Is the Ladder True or False?	Is Output Pilot Light On?
Open			
Closed			
NORMALLY CLOSED PUSH BUTTON ADDRESS I:1/0			
Push Button	—I / I— True or False?	Is the Ladder True or False?	Is Output Pilot Light On?
Open			
Closed			

Figure 12-7

52. _____ Fill in the table in Figure 12-8 as you observe how the input instructions react to the toggle switches' open or closed conditions.

NORMALLY OPEN PUSH BUTTON ADDRESS I:1/1			
Push Button	—I I— True or False?	Is Input Status File Bit True or False?	Is the Output Status File Bit True or False?
Open			
Closed			
NORMALLY CLOSED PUSH BUTTON ADDRESS I:1/1			
Push Button	—I / I— True or False?	Is Input Status File Bit True or False?	Is the Output Status File Bit True or False?
Open			
Closed			

Figure 12-8

What Have We Learned?

It is important to understand the relationship between the physical input device, such as a normally closed push button, and the ladder program instruction necessary to achieve the correct logic. Keep in mind that the physical input device and its programmed ladder instruction representation are programmed independently of how the input device is physically wired. As a result, the status of a normally open push button can be tested not only with a normally open instruction, but with a normally closed instruction. In the course of completing this lab, you experimented with the different combinations of normally open and normally closed instructions and how they interact with different combinations of physical input conditions—with a normally open and a normally closed physical input device.

The question is, What logic needs to be programmed to make my ladder program instruction true at the desired time? As an example, take a normally closed push button. When does the instruction associated with the physical input need to be true? If the normally closed push button results in a true ladder instruction when its in its normally closed state, this exercise and the table you developed in Figure 12-8 should lead you to program a normally open ladder program instruction. If on the other hand, the normally closed push button is to result in a true instruction when in the activated, or physically open, state, a normally closed ladder program instruction would be selected.

In a second example, suppose you have a normally open mechanical limit switch being held closed as its normal condition. You need to know when the object holding the switch moves away and releases the switch. What instruction would you program on your PLC ladder so the rung is true only when the object holding the limit switch closed moves out of the way? Refer to the tables you developed above. You should choose a normally closed programming instruction.

Always remember that PLC instructions change from true to false, or false to true. Accordingly, a normally open PLC instruction does not change to a normally closed instruction, nor does a normally closed instruction change to a normally open instruction. Even though electromechanical relay contacts physically open or close, PLC ladder instructions test the input status data table location associated with the real-world input device for a 1 or 0 logical level.

PART III

For this application, we will experiment with a normally open and a normally closed push button replacing our toggle switch inputs.

1. _____ Remove power to your PLC. Install a normally closed, momentary push button for I:1/3, and a normally open, momentary push button for I:1/2.
2. _____ Using the program editing skills you acquired in this exercise, edit the ladder program so it looks like Figure 12-9 on the next page.
3. _____ Go on-line and down load your newly edited program.
4. _____ Put the PLC processor in run mode.
5. _____ As your PLC program is running, Experiment with rungs four and five to determine which combinations of ON and OFF signals from the two input push buttons make each ladder rung true.
6. _____ Fill in the tables below as you collect data from your experiments. Fill in the table in Figure 12-10 (next page) for rung four's data.

What Have We Learned from Rung Four?

1. The normally open push button, I:1/2, only inputs a signal into the input module and input status file when its button is depressed.
2. The normally closed push button, I:1/3, inputs a signal into the input module and input status

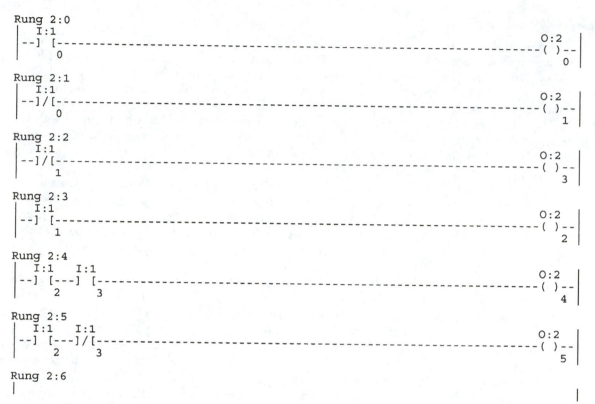

Figure 12-9 Edited program for this exercise.

RUNG ZERO						
Push Buttons: Pressed?		**On Signal to Input Module ?**		**Input Data File Bit**		**Output O:2/4**
I:1/2 N.O.	**I:1/3 N.C.**	**I:1/2**	**I:1/3**	**I:1/2**	**I:1/3**	**True or False?**
No	No					
No	Yes					
Yes	No					
Yes	Yes					

Figure 12-10 Push button behavior in relation to PLC I/O and program.

file when the button is not pressed. As a result, the XIC instruction associated with a normally closed input will be true all the time until the push button is pressed. As an example, a normally closed *stop* push button would be configured and programmed in this manner so as to provide logical continuity at all times except when the button is pressed to input a stop signal.

3. _____ Fill in the table in Figure 12-11 for rung five's data.

RUNG FIVE						
Push Buttons: Pressed ?		**On signal to input module ?**		**Input Data File Bit**		**O:2/5 Output**
I:1/2 N. O.	**I:1/3 N.C.**	**I:1/2**	**I:1/3**	**I:1/2**	**I:1/3**	**True or False?**
No	No					
No	Yes					
Yes	No					
Yes	Yes					

Figure 12-11 Push button behavior in relation to PLC I/O and program.

13

LATCHING
INSTRUCTIONS

This exercise will coach you in editing, or changing, your *Ladder_1* program, incorporating latching and one-shot instructions.

First we will work with the output latch (OTL) and the output unlatch (OTU) instructions. The output latch instruction is used to latch an output bit, either a real output or an internal bit. When a bit is latched, the bit is turned ON and latched until the bit is reset by the unlatch instruction. The *latch* instruction is only used to latch an output bit ON. The *output unlatch* is only used to unlatch an output bit that has previously been latched. As a result, you will need to use both of these instructions together as a pair, to latch an output one ladder rung while the unlatch instruction on a different rung with the same output address will unlatch the output bit. Our renamed *Ladder_1* program will be edited to look the same as Figure 13-1.

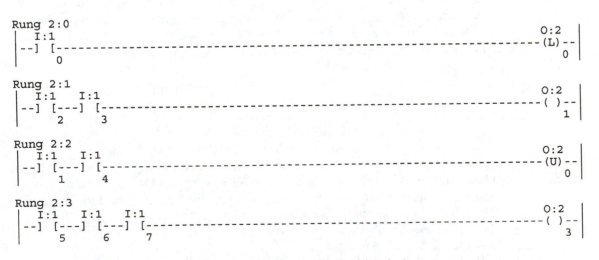

Figure 13-1 Edited *Ladder_1* program, now named *Latch_1*.

THE LAB

1. _____ Copy the *Ladder_1* file from your student floppy disk to your personal computer's hard drive.

2. _____ Rename the *Ladder_1* file *Latch_1*.

3. _____ Press *Escape* twice.

4. _____ Select file by pressing F4, *Change File*.

5. _____ Use arrow keys to select *Latch_1* file.

6. _____ Press F9, *Save To File*.

7. _____ Press F1, *Offline Programming / Documentation*.

8. _____ Press F8, *Monitor File*.

9. _____ Move the cursor to the O:2/0 instruction. We will edit this instruction and change it to an output latch instruction with the same address.

10. _____ Press F10, *Edit*.

11. _____ F5, *Modify Rung*.

12. _____ Press F5, *Modify Instruction*.

13. _____ Press F1, *Bit* to select the replacement instruction.

14. _____ Press F4, *OTL* (output latch). This will change the OTE output instruction to the output latch (OTL) instruction.

15. _____ The message, "Enter Bit Address > O:2/0," provides the opportunity to modify the instruction's address. *Do not change the address*. Press *Enter* to accept the current address.

16. _____ Press *Enter* or *Escape* to get to the *Accept Rung* function.

17. _____ Press F10 to *Accept Rung*.

18. _____ In order for the latch instruction to be unlatched, we must have a separate unlatch instruction with the same output bit address as the latch instruction programmed somewhere in our program. We will program an unlatch O:2/0 instruction so we can unlatch O:2/0 once it is latched. Move the cursor to *O:2/2 instruction*.

19. _____ Press F5 to *Modify Rung*.

20. _____ Press F5 to *Modify the instruction*.

21. _____ To change the current OTE O:2/2 instruction to an unlatch instruction (OTU) with the address O:2/0, press F1, *Bit*.

22. _____ Press F5, *OTU* (Output Unlatch).

23. _____ Where it says, "Enter Bit Address > O:2/2," the cursor should be flashing just to the right of the current address. To change this address to O:2/0, press the backspace key one time to delete the output point reference. Type in the new output point reference of 0. You should have changed the original address from O:2/2 to O:2/0.

24. _____ Press *Enter* to accept the address.

25. _____ Press *Escape* or *Enter* to go to the *Accept Rung* function screen.

26. _____ Press F10 to *Accept Rung*.

27. _____ Notice that the latch and unlatch instructions both have the same address. The latch instruction, on rung zero, is controlled by I:1/0. The output unlatch instruction could be programmed on any other rung. In our exercise, we choose rung two. We could have placed the unlatch instruction on any rung after the rung with the latch instruction. Placement of the unlatch instruction would be decided by determining whether there are a series of rungs containing instructions that should be executed before the unlatch instruction would be activated.

28. _____ Go online and download your program.

29. _____ Put the PLC into run mode.

30. _____ Turn toggle switch, input address I:1/0 ON and then OFF.

31. _____ Output O:2/0 should come ON and stay on, even though the input is now OFF. The

output is said to be *latched ON*. The latched output will remain latched ON until the *unlatch* instruction with the same address becomes true.

32. _____ With output O:2/0 latched ON, and input I:1/0 in the OFF condition, turn off power to the PLC.

33. _____ After a few seconds, when all the LEDs on the processor have gone out, power up the PLC.

34. _____ When the PLC has completed its power up self-test sequence, notice that the latched output O:2/0 comes back ON. A latched output will return to its last state before power loss. This is called a *retentive* output.

35. _____ Energize inputs I:1/1 and I:1/4. The unlatch instruction should be true.

36. _____ The latch instruction should be false. Output O:2/0 should be OFF.

37. _____ With the unlatch instruction still true, try to latch output O:2/0 ON.

38. _____ You should discover that it is not possible to latch an output bit ON when its unlatch instruction is true.

PROGRAMMING LATCHING OR SEALING THREE-WIRE CONTROL LOGIC

A momentary start-stop push button station is typically used to start and stop many industrial processes. The next part of this exercise will connect and program a typical three-wire start-and-stop push-button station.

WHAT IS NEEDED TO COMPLETE THIS LAB EXERCISE

1. Your PLC.
2. APS Programming software.
3. A start-stop push-button station. The start push button should be momentary, normally open. The stop push button should be momentary, normally closed. If a start-stop station is not available, use separate push buttons. We are going to add rung three to our previous latching ladder program. Figure 13-2 illustrates what we are going to do for this exercise.

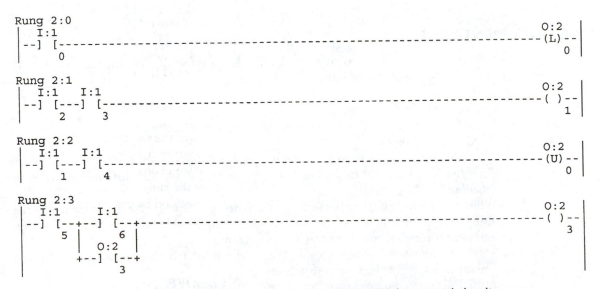

Figure 13-2 Three-wire sealing logic for a typical three-wire control circuit.

THE LAB

Disconnect power from your PLC. Connect the normally closed stop push button to PLC input screw terminal address I:1/5. Connect the normally open start push button to PLC input screw terminal address I:1/6.

1. _____ Go off-line and edit your ladder program.
2. _____ F10, *Edit*.
3. _____ F3, *Offline Disk*.
4. _____ Use the arrow keys to go to input I:1/7.
5. _____ Press F5, *Modify Rung* to edit this ladder rung.
6. _____ Press F6, *Delete Instruction*. I:1/7 should disappear.
7. _____ We will now add a branch around the I:1/6 instruction. Use the arrow keys to move the cursor to highlight I:1/6.
8. _____ Press F1, *Branch*.
9. _____ Press F4, *Insert branch*.
10. _____ To branch around I:1/6, select target C, F3.
11. _____ Press F4, *Insert instruction*.
12. _____ To add the XIC instruction, Press F1, *Bit*.
13. _____ Press F1, *XIC*.
14. _____ Enter Bit Address > *O:2/3*.

      Yes, we want to enter the output address to this input instruction. When we push the *Start* momentary push button, the output will be true or energized. With the output bit O:2/3 energized the input instruction associated with address O:2/3—which is the same bit, only referenced to an input XIC instruction—will also become true. With the parallel branch true, the operator can release the momentary *Start* push button and the parallel branch will seal or latch the circuit. The circuit will remain latched until logical continuity is lost either as a result of power loss making I:1/5 input a false input signal, or an operator pressing *Stop* push button I:1/5, breaking logical continuity.

15. _____ Press *Enter* to accept address.
16. _____ Press *Escape* or *Enter* to go to the *Accept Rung* function.
17. _____ Press F10 to *Accept Rung*.

ADDING DOCUMENTATION

We will add some documentation text to help identify the purpose of each instruction. A rung comment will be entered to explain the function of the rung.

1. _____ Cursor to the I:1/5 instruction.
2. _____ Press *Escape* to get to the screen with the documentation function.
3. _____ Press F5, *Document*.
4. _____ To add identifying text to this instruction, press F2, *Instruction Comment*.
5. _____ Type the following text in the box that appears on the screen: *Normally closed stop push button.*
6. _____ When completed entering the text, press F8, *Accept / Exit* to save the text and exit the Entering Documentation screen. If text fails to display, skip ahead to the next section and read through displaying or suppressing comments.
7. _____ The text you entered should appear directly above the I:1/5 instruction on the ladder.
8. _____ Cursor to highlight the I:1/6 instruction.
9. _____ Press F2, *Instruction comment* to add text to this instruction.
10. _____ Enter the following text: *Normally open start push button.*

11. _____ Press F8 to accept the instruction comment and exit the enter instruction comment screen.
12. _____ The comment should be directly above the I:1/6 instruction.
13. _____ Cursor to the parallel branch and the XIC O:2/3 instruction.
14. _____ Press F2, *Instruction Comment*.
15. _____ Add the following text: *Hold in or latch*.
16. _____ Press F8 to *Accept and Exit*.
17. _____ Next we will add a general comment to describe the entire rung. This is called a rung comment. Press F1, *Rung Comment*.
18. _____ An area is opened directly above the rung. Enter text for the rung comment as illustrated in the figure below:

```
Rung 2:0
|   I:1                                                                    O:2  |
|--] [-------------------------------------------------------------- (L)-- |
|    0                                                                      0   |

Rung 2:1
|   I:1    I:1                                                             O:2  |
|--] [---] [-------------------------------------------------------- ( )-- |
|    2      3                                                              1   |

Rung 2:2
|   I:1    I:1                                                             O:2  |
|--] [---] [-------------------------------------------------------- (U)-- |
|    1      4                                                              0   |

Rung 2:3
This is a latching rung where the normally-open start push button energizes
output O:2/3. Being a momentary push button, the parallel branch O:2/3 XIC
instruction will also energize when output O:2/3 energizes. Input instruction
I:2/3 will provide an alternate path for logical continuity when the operator
releases the start push button.  Opening the rung by pressing the stop push
button, XIC I:1/5 will disrupt continuity and unlatch the rung and output O:2/3.
|  Normally      Normally                                                      |
|  Closed        Open Start                                                    |
|  Stop Push     Push                                                          |
|  Button        Button                                                        |
|    I:1           I:1                                                    O:2   |
|----] [-----+----] [-----+------------------------------------------( )--   |
|    5       |     6       |                                              3     |
|            | Hold in or  |                                                   |
|            | latching    |                                                   |
|            | instruct-   |                                                   |
|            | ion         |                                                   |
|            |    O:2      |                                                   |
|            +----] [-----+                                                   |
|                 3                                                            |
```

Figure 13-3 *Latch_3* program with rung comment.

19. _____ When finished entering the rung comment, press F8, *Accept / Exit*.
20. _____ Press F10, *Save Documentation*. This will save the documentation to the hard drive.

DISPLAYING OR SUPPRESSING COMMENTS

Your personal computer screen can be set up to display or suppress documentation comments. This selection is found under the configure display function key. Let's look into configuring our display to either display or suppress comments.

1. _____ Press *Escape.*
2. _____ Press F2, *Configure Display.* This will take you to the function keys that allow the operator to either view or suppress documentation.

The following keys are used to either display or suppress comments:

F6: This function key either displays or suppresses cross reference (XREF) information. This feature will display a cross-reference showing where else a specific address is used in your ladder program. If you press F6 until its text reads "Suppress XREF," there will be one cross-reference displayed directly below the parallel branch on rung three. Directly under the XIC O:2/3 instruction on rung three will be displayed (2:3). This is the cross-reference. The cross-reference tells us two things: a. In what ladder program file this address is used again, and b. At which rung in that program file this address is used again. The cross-reference (2:3) states program file 2, rung 3, has another occurrence of this instruction address. Since we are already in program file 2 (refer to your text for information on program files) on rung 3, this cross-reference is referring to the OTE instruction on this rung. O:2/3 on this rung has the same address as the parallel XIC instruction.

Looking at the personal computer screen: the current file and rung reference is displayed directly to the left of the flashing cursor near the lower left corner of the screen. Notice the reference: File 2, rung 3. This should be next to the flashing cursor. You may need to use the arrow keys to be on rung 3. If you are on another program rung, that rung number will be displayed. Use the up and down arrow keys to move between rungs, noticing how the rung reference changes.

F7: Suppress Rung Comments / Display Rung Comments: If rung comments are being displayed, the F7 selection will say, "Suppress rung comments." Push F7 to suppress rung comments. Try it.

When rung comments are suppressed, the F7 selection shows, "Display rung comments." To display rung comments, press F7.

F8, Suppress Instruction Comments / Display Instruction Comments: This key works similarly to the F7 key for rung comments. To suppress instruction comments press F8 until the function text says, "Display instruction comments." To display instruction comments, push F8.

F9, Suppress Instruction Symbols / Display Instruction Symbols: This key works similarly to the F7 and F8 keys. To suppress instruction symbols press F9 until the function text says, "Display instruction symbols." To display instruction comments simply push F9.

F10, Save Configuration:

3. _____ Press F10, *Save configuration* after you have selected you desired screen configuration.
4. _____ Press *Escape.*
5. _____ F3, *Exit.*
6. _____ F2, *Save.*
7. _____ Save program to your student floppy.

A NOTE REGARDING LADDER PROGRAM DOCUMENTATION

All program documentation is stored on your personal computer's hard drive. Program documentation is not downloaded to your PLC processor. This means that if you upload the PLC processor's ladder program to another personal computer, no documentation comments will upload. If you upload to the personal computer that has the documentation files on its hard drive, documentation can be accessed; if uploading to another personal computer, where those documentation files are not resident, there will be no documentation.

14

THE ONE-SHOT INSTRUCTION

This lab will edit Lab Thirteen's latch ladder program and add a rung containing a one-shot instruction. The next rung we will develop will contain a one-shot rising (OSR) instruction. The new rung four, illustrated in Figure 14-1 on the following page, will be incorporated into our current ladder program.

When input I:1/7 goes from false to true, the OSR instruction will allow the output O:2/4 to turn on for only one processor scan: it will transition from false to true, be true for one processor scan, and transition back to false. O:2/4 will remain false until input I:1/7 transitions from false to true again. It makes no difference if I:1/7 is momentary or maintained. The OSR instruction will allow the output to be true for one scan even if I:1/7 stays on continuously. The OSR instruction is called a one-shot-rising instruction because the instruction must see a rising false-to-true transition from associated input logic before the output will be triggered. Let's enter our new rung.

THE LAB

1. _____ Open Lab Exercise Thirteen's ladder program and place cursor on rung three.
2. _____ Press F10, *Edit*.
3. _____ Press F3, *Append Rung*. This adds rung four, which we will program with the OSR instruction.
4. _____ Press F4, *Insert Instruction*.
5. _____ Press F1, *Bit*.
6. _____ Press F1, *XIC*.
7. _____ Enter bit address > *I:1/7*.
8. _____ Press *Enter* to accept address.
9. _____ Press F1, *Bit*.
10. _____ To add the one-shot instruction in series with I:1/7; press F6, *OSR*.
11. _____ Enter a bit address from bit file 3. (The OSR instruction does not use a real world I/O address. Use a bit address such as B3:0/1. This internal bit address is needed to track

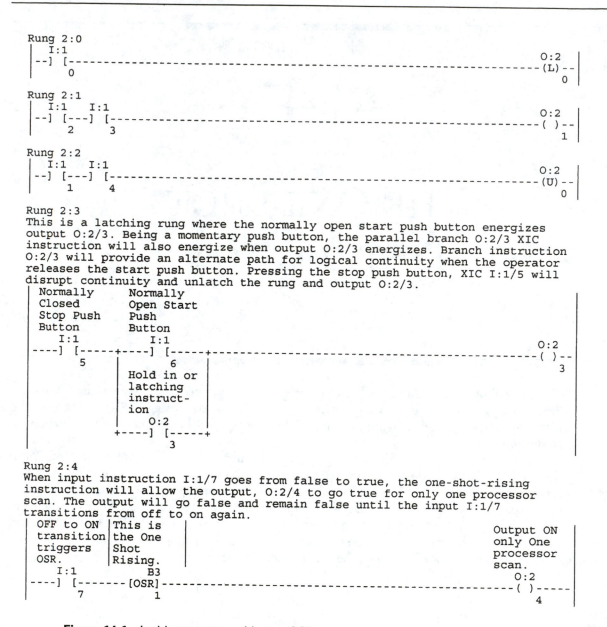

Figure 14-1 Ladder program with new OSR rung.

the status of the OSR instruction only. This address breaks down to bit file 3, word 0, bit 1.)

12. _____ Press *Enter* to accept address.
13. _____ Press F1, *Bit*.
14. _____ Press F3, *OTE* for the rung's output instruction.
15. _____ Enter bit address > *O:2/4*.
16. _____ Press *Enter*.
17. _____ Press *Escape* or *Enter* to go to the Accept Rung function screen.
18. _____ Press F10, *Accept Rung*.
19. _____ Press *Escape* to delete the extra rung appended for you.
20. _____ Move the cursor back to rung four.

21. _____ Press *Escape* to get to the documentation function key.
22. _____ Press F5 to go to the Documentation Selection functions screen.
22. _____ Add documentation as in Figure 14-1.
23. _____ When finished entering documentation, *Save* and *Go On line*.
24. _____ Put processor in Run mode and test your program.
25. _____ A note regarding program operation: You will not see much when the rung with the OSR instruction is executed. This is because the output will only be true for one scan.
26. _____ When finished, save the program to your student floppy.
27. _____ Return to the main menu.

TYPICAL APPLICATIONS FOR THE ONE-SHOT INSTRUCTION

Below are three applications where the one-shot instruction would be typically used. Figure 14-2 illustrates a one-shot controlling a BCD conversion instruction, the TOD. The TOD instruction is the SLC 500 integer-to-BCD conversion instruction. An example of a ladder rung containing the TOD instruction is illustrated in Figure 14-2.

```
Rung 2:0
  I:1    B3                                              +TOD-------------+
 --] [--- [OSR] --------------------------------------+TO BCD          +-
    0      6                                           | Source   T4:0.ACC|
                                                       |                0|
                                                       | Dest       O:2.0|
                                                       |             0000|
                                                       +----------------+
```

Figure 14-2 One-shot used in conjunction with a to-BCD conversion instruction.

When the input instruction transitions from false to true, the OSR will trigger the TOD instruction once. The TOD instruction will take the value of T4:0.ACC, convert it to BCD, and output it to the destination output O:2. The one-shot instruction will allow a stable display if updated at a predetermined frequency. It is easier to read a stable display that is updated every few seconds, than a constantly changing display.

A second use for a one-shot instruction is in conjunction with a message instruction. The message instruction is used to send information from one SLC 500 processor to another over the Data Highway 485 Network. When triggered, the message display sends the data only once, as a result of the one-shot instruction. Figure 14-3 illustrates the message instruction used in conjunction with the OSR instruction.

A third possible use for a one-shot instruction is to provide one input pulse for a closure of a momentary start push button. No matter how long the button is held depressed, only one pulse

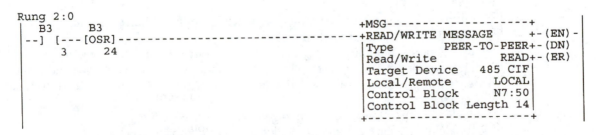

```
Rung 2:0
  B3     B3                                         +MSG------------------+
 --] [--- [OSR] -----------------------------------+READ/WRITE MESSAGE      +-(EN)-
    3      24                                       | Type        PEER-TO-PEER+-(DN)
                                                    | Read/Write         READ+-(ER)
                                                    | Target Device   485 CIF|
                                                    | Local/Remote      LOCAL|
                                                    | Control Block     N7:50|
                                                    | Control Block Length 14|
                                                    +---------------------+
```

Figure 14-3 One-shot instruction used to control a message instruction.

```
‖ ┌─────────┬─────────┐
‖ │Push-on /│One-shot │
‖ │Push-off │so push  │
‖ │momentary│button   │
‖ │push     │will input│
‖ │button.  │one pulse.│
‖   I:1         B3                                                    B3  │
‖ ----] [------- [OSR] -------------------------------------------------( )--│
‖     0          0                                                     2  │
```

Figure 14-4 One-shot used in conjunction with a momentary push button.

will be seen by the PLC as an input. Figure 14-4 illustrates part of the ladder logic used in conjunction with the momentary push button.

PRINTING A HARD COPY OF YOUR PROGRAM

When a hard copy of a ladder program and the associated data is desired, reports need to be generated and then the reports can be printed out to your printer.

THE LAB

(Your instructor will show you how to interface your personal computer with APS Software to an available printer.) Before we can print our program, reports must be created. It is the reports that will be printed.

1. _____ Start at the main menu screen.
2. _____ Press F3, *Off line programming / Documentation*.
3. _____ To create reports for printing, press F6, *Create Reports*.
4. _____ Two windows should pop up on your screen. The right box provides a selection as to what is to be printed. To view or change, push F5, *Report Options*. When you push F5, the screen changes and the function key selections change to reflect your program listing options.
5. _____ If you make any changes, press F9, to *Save Options to file*.
6. _____ Press *Escape* to return to the previous screen.
7. _____ The box on the right entitled *Documentation* selects what reports will be generated.
8. _____ Press F2 to select all reports. A star should appear next to each report selection.
9. _____ Press F4 to reset all report options. The stars should disappear. There are now no reports selected.
10. _____ Pressing the down arrow moves from one report to the next. Options will be displayed with each selection. For this exercise leave all options as they are. Refer to the Allen-Bradley *APS Software User Manual* for additional information on printing reports.
11. _____ Use the arrow keys to select the reports you wish to generate. When each is highlighted, press F3, *Toggle Report* to select it.
    F6, *Option* should already be configured for your printer. This contains printer configuration options, along with other general options such as paper size and print mode. If you pressed F6 to view the general options, press *Escape* to return to the previous screen.
12. _____ After selecting the reports you want to generate, press F8, *Title*.
13. _____ Use the delete key on the keyboard to delete old title information.
14. _____ Enter your name and the program name, *Latch_1*, as the new title.

15. _____ Press F9 to save to file.
16. _____ Press *Escape* to return.
17. _____ Press *Enter* to generate reports.
18. _____ When reports are generated, the message, "Press a key to continue," is displayed. Press *Enter* to return to the Program Directory Screen.
19. _____ Press F3 to return to the main menu.
20. _____ When you are ready to print, press F8, *Print Reports*.
21. _____ Use the arrow keys to select the *Latch_1* processor file.
22. _____ Press *Enter*.
23. _____ Press F2 to select all reports, or use the arrow keys and F3 toggle select to select individual reports.
24. _____ Print the reports by pressing F7.
25. _____ When printing is complete, press *Escape* to return to the main menu screen.

15

USING INTERNAL BITS TO MAKE A PUSH-ON/PUSH-OFF PUSH BUTTON

For this exercise we need to interface a single momentary push button so that when an operator presses the button once the process starts; when the button is pushed a second time, the process stops. The exercise will look into developing the ladder program to make this push button work as dictated by the application.

Developing the ladder logic to accomplish this will include a number of instructions needed to set up the proper operating sequence, which will not be physical, outside input points or output points. These instructions will be assigned internal memory addresses in a file called the bit file. The bit file stores single-bit addresses for storage of the ON or OFF state of the instructions associated with the assigned address. When real-world references are not needed, but instructions like XIO, XIC, OSR, OTL, OTU, and OTE need to set up programmed logic which will only be used internally to control other logic, internal bits and the bit file are used. Some manufacturers call these bit references; others refer to them as internal coils, internal relays or internal bits. In most cases these names are synonyms.

ADDITIONAL MATERIALS NEEDED

One normally open push button is to be wired to PLC input I:1/1 if you have a modular PLC. Connect the push button to input address I:0/1 if you have a fixed SLC 500 or a MicroLogix 1000.

INTRODUCTION

When the operator presses the push button, the PLC needs to see an input pulse for just one scan. If, as a result of contact bounce, multiple input pulses are seen by the PLC, these multiple signals could be interpreted as multiple start or stop signals, and possibly cause unpredictable operation.

To insure that only one input signal is input to the PLC, the momentary push button's input instruction will be programmed in series with a one-shot instruction. We will develop internal logic, or instructions, to accomplish the task at hand. The internal logic we will develop will establish how the push button's input signal is to be handled by the PLC. Internal logic will consist of XIC and XIO instructions that have internal memory addresses rather than real-world input or output addresses. Internal memory addresses are stored in a file similar to the input status or output status data file. Whereas output status data was stored in file zero, input status data stored in file one, and processor status data stored in file two, internal bit data will be stored in data file three, the bit file. Bit file three is the default bit file created when your processor creates a processor file. Any unused data file above file ten can be configured as an additional bit file.

We will need to add these internal instructions between the push button's input instruction and the real-world output device's output instruction. Bit file three is typically used for internal bits, shift registers and sequencers. A bit file is made up of 255 one-word elements. One element is a word containing sixteen bits. Each bit is individually addressable. There are a total of 4096 (Bits 0 through 4095) bits that make up a bit file. Figure 15-1 illustrates an example of a bit file.

15	14	13	12	11	10	9	6	7	6	5	4	3	2	1	0	←Bit
0	0	0	0	0	0	0	0	0	0	0	0	0	0	0	0	B3:0
0	0	0	0	0	0	0	0	0	0	0	0	0	0	0	0	B3:1
0	0	0	0	0	0	0	0	0	0	0	0	0	0	0	0	B3:2
0	0	0	0	0	0	0	0	0	0	0	0	0	0	0	0	B3:3
0	0	0	0	0	0	0	0	0	0	0	0	0	0	0	0	B3:4
0	0	0	0	0	0	0	0	0	0	0	0	0	0	0	0	B3:5
0	0	0	0	0	0	0	0	0	0	0	0	1	0	0	0	B3:5

Bit 100 in file

Figure 15-1 Representation of a bit file.

BIT FILE ADDRESSING FORMAT

The bit file addressing format is very similar to input status file and output status file addressing. A typical bit file address is illustrated in Figure 15-2.

Alternate Bit Addressing Format

Bit file bits can also be addressed by identifying the file type, file number and bit number. Since there are 4096 bits in a bit file, and the first bit is zero, the bit range is from 0 to 4095. Addressing by bit number format is illustrated in Figure 15-3.

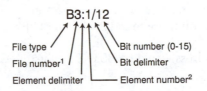

B3:1/12

File type
File number[1]
Element delimiter
Bit number (0-15)
Bit delimiter
Element number[2]

[1]Bit file 3 is the default bit file. Any unused data file between 10 and 255 can be a bit file.
[2]Bit file elelments are single sixteen bit words. There are 256 elements (16 bit words) per bit file.

Figure 15-2 Addressing for bit file 3, word or element 1, bit 12.

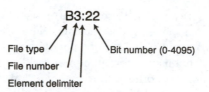

Figure 15-3 Bit file address for bit file 3, bit 22.

The example address from Figure 15-3, B3:22, can easily be converted to the element-and-bit addressing format by first determining how many 16-bit words are used. Divide the number of bits by 16 (bits per element) to get the number of elements. Add one to the number of elements to include the element containing the remainder bits. This will give you the total number of elements, which is also going to be the element number in your address.

The element containing the remainder bits should contain fewer than 16 bits. Determine the bit position in the remaining element to get the bit number in your address. Remember to count bit zero in this element. Two examples are illustrated below:

Example One: Determine element and bit address for B3:22.

1. Bit Number is 22; 22 divided by 16 equals 1 with a remainder of 6. One element, element zero, plus the element containing the remainder bits, puts the element address at B3:element 1.
2. The six remainder bits are found in element 1. Counting from bit zero, the sixth bit in element one is bit five.
3. The converted address would be B3:1/5, as illustrated in Figure 15-4.

15	14	13	12	11	10	9	8	7	6	5	4	3	2	1	0	←Bit
0	0	0	0	0	0	0	0	0	0	0	0	0	0	0	0	B3:0
0	0	0	0	0	0	0	0	0	0	1	0	0	0	0	0	B3:1

Bit 100 in file

Figure 15-4 Bit file address for B3:22 bit file position.

Example Two: Determine element and bit address for B3:100.

1. Bit Number is 100; 100 divided by 16 equals 6 with a remainder of 4. From element zero, the sixth element in our bit file is element five. Adding the element containing the remainder bits, the element address portion of the converted address is B3:element 6.
2. The four remainder bits are found in element 6. From bit zero, the fourth bit in element six is bit three.
3. The converted address would be B3:6/3, as illustrated in Figure 15-5.

CREATING AN ADDITIONAL BIT FILE

Even though bit file 3 is the default, automatically created when a new processor file is created, additional bit files may be created for additional, or separate, bit storage. This exercise will show you how to create an additional bit file, bit file 10. We will then create a ladder program containing bit instructions.

15	14	13	12	11	10	9	8	7	6	5	4	3	2	1	0	←Bit
0	0	0	0	0	0	0	0	0	0	0	0	0	0	0	0	B3:0
0	0	0	0	0	0	0	0	0	0	0	0	0	0	0	0	B3:1
0	0	0	0	0	0	0	0	0	0	0	0	0	0	0	0	B3:2
0	0	0	0	0	0	0	0	0	0	0	0	0	0	0	0	B3:3
0	0	0	0	0	0	0	0	0	0	0	0	0	0	0	0	B3:4
0	0	0	0	0	0	0	0	0	0	0	0	0	0	0	0	B3:5
0	0	0	0	0	0	0	0	0	0	0	0	1	0	0	0	B3:6

Bit 100 in file

Figure 15-5   Alternate address for B3:100, B3:6/3.

1. _____ Copy the *Begin* file from your student floppy to your personal computer's hard drive.
2. _____ Rename the *Begin* processor file, *Push_On*.
3. _____ Press *Escape* twice to get to the *Program Directory for Processor* screen.
4. _____ PressF4, *Change file*.
5. _____ Press F3, *Off line programming / Documentation*.
6. _____ Press F4, *Change File*. Use the arrow keys to select *Push_on* file.
7. _____ Press F9, *Save to file*.
8. _____ Press F1, *Off Line programming / Documentation*.
9. _____ If you get the warning: "Processor program name and DOS file name mismatch; Resave file. Press any key to continue," Press F2, *Save*.
10. _____ Press F8, *Yes,* to accept defaults.
11. _____ Press F8, *Monitor File*.
12. _____ Press F7, *General Utility*.
13. _____ We are going to view our memory map. Press F1, *Memory Map*. Figure 15-6 illustrates the default memory map.

```
                      DATA TABLE MAP

    FILE        TYPE          LAST ADDRESS     ELEMENTS    WORDS     FILE
                                                                    PROTECTION

     0       O output           O:6              4          4       STATIC
     1       I input            I:4              2          2
     2       S status           S:82             83         83
     3       B binary or bit    B3/15            1          1
     4       T timer                             0          0
     5       C counter                           0          0
     6       R control                           0          0
     7       N integer                           0          0
     8       F float                             0          0
     9         unused                            0          0

                  PROCESSOR MEMORY LAYOUT

        190 data words of memory used in  11 data table files
         16 instruction words of memory used in 3 program files
      12272 instruction words of unused memory available


    Press a key or enter file number

    offline              SLC 5/03
    File PUSH-ON
                         CREATE    DELETE                   DATA
                         DT FILE   DT FILE                  PROTECT
                           F6        F7                     F10
```

Figure 15-6   SLC 500 default memory map.

14. _____ We are going to create an additional bit file, bit file 10. To create a new data table file, press F6, *Create Data Table File*.

15. _____ Our bit file needs to have 100 elements, or words (elements 0-99). At "Enter Address To Create," enter *B10:99*. This will create data file 10, a bit file, with 100 elements.

16. _____ Press *Enter* to accept.

17. _____ To view Bit file 10, under the message, "Press a key or enter a file number", enter *10* for file number ten. Press *Enter*. Bit file ten will be displayed on page two of the data table map screen. Your monitor should appear like Figure 15-7.

```
                            DATA TABLE MAP

    FILE    TYPE              LAST ADDRESS    ELEMENTS   WORDS    FILE
                                                                 PROTECTION
    10      B binary or bit   B10/1599          100       100




                        PROCESSOR MEMORY LAYOUT

          190 data words of memory used in  11 data table files
           16 instruction words of memory used in 3 program files
        12272 instruction words of unused memory available


    Press a key or enter file number

    offline                 SLC 5/03
    File PUSH-ON
                            CREATE    DELETE                    DATA
                            DT FILE   DT FILE                   PROTECT
                            F6        F7                        F10
```

Figure 15-7   Bit file B10 displayed on page 2 of the memory map.

18. _____ When you are done viewing the data table map, press *Escape* twice to return to the screen where you can select F10, *Edit*.

DEVELOPING A LADDER PROGRAM FOR A PUSH-ON/PUSH-OFF APPLICATION

The following steps lead you through the development of this application's program. Figure 15-8 on the next page is the ladder program for this exercise.

1. _____ Press F10, *Edit*.

2. _____ Press F4, *Insert Rung*.

3. _____ Press F4, *Insert Instruction*.

4. _____ Press F1, *Bit*.

5. _____ Press F1, *XIC*. This is your momentary normally open push button instruction.

6. _____ Enter Bit Address > *I:1/0*.

7. _____ Press *Enter* to accept address.

8. _____ Press F1, *Bit* to go to the screen with the OSR instruction.

9. _____ Press F6 to select the one-shot (OSR instruction).

10. _____ Enter bit address > *B3:0/0*. This is bit file three, word or element zero, bit zero. Remember, the OSR instruction is addressed with a bit address.

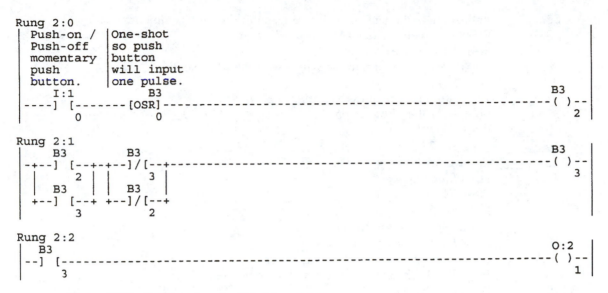

Figure 15-8 Push-on / push-off ladder program.

11. _____ Press *Enter* to accept the address.
12. _____ Press F1, *Bit* so we can select the output instruction.
13. _____ Press F3, *OTE*.

We have to develop some internal logic in order to tell the PLC processor how to react when it sees an input from I:1/0. Remember, the OSR instruction will only allow a one scan pulse (one-shot) no matter how long the I/O push button is depressed.

Since we are not ready to control our hardware output device yet, we will use internal addresses to control the instructions we are about to program.

14. _____ Enter Bit address > *B3:0/2*. This is bit file 3, element 0, bit 1. This is the next available bit address. This internal bit will be true only when the rung's input logic is true. Since this is not a real output address, this bit is only represented in internal memory, of bit file 3.
15. _____ Press *Escape* or *Enter* to go to the Accept Rung screen selection.
16. _____ Press F10 to *Accept Rung*.
17. _____ Add documentation to the rung just completed. Press *Escape* to go to the Document Function screen.
18. _____ Cursor to I:1/0 instruction.
19. _____ Press F5, *Document*.
20. _____ We will add an instruction comment. Press F2, *Instruction Comment*.
21. _____ Add instruction comment for this instruction as illustrated in Figure 15-8.
22. _____ Press F8, *Accept / Exit*.
23. _____ Cursor to the OSR Instruction.
24. _____ Press F2, *Instruction Comment*.
25. _____ Add instruction comment for this instruction as illustrated.
26. _____ When completed, press F8, *Accept / Exit*.
27. _____ Cursor to the OTE instruction.
28 _____ Press F2, *Instruction Comment*.
29. _____ Add the instruction comment for this instruction as illustrated.
30. _____ When completed, press F8, *Accept / Exit*.
31. _____ Entering a rung comment. Press F1, *Rung Comment*.

32. _____ In the area that opens above the rung, enter the rung comment.
33. _____ When completed, Press F8, *Accept / Exit*.
34. _____ Press F10 to save documentation.
35. _____ Press Escape to return to the screen with the edit selection.
36. _____ Press F10, *Edit*.
37. _____ To create rung one, press F3, *Append Rung*.
38. _____ Press F4, *Insert Instruction*.
39. _____ Press F1, *Bit*.
40. _____ Press F1, *XIC*.
41. _____ Enter Bit Address > *B3:0/2*.
42. _____ Press *Enter* to Accept.
43. _____ Press F1, *Bit*.
44. _____ Press F2, *XIO*.
45. _____ Enter Bit Address > *B3:0/3*.
46. _____ Press *Enter* to accept address.
47. _____ Use the arrow keys to move rung cursor to the left power rail.
48. _____ Press *Escape*. This will take you to the branch selection.
49. _____ Press F1, *Branch*.
50. _____ Press F4, *Insert Branch*.
51. _____ Since you are currently at target *A*, and want to branch around B3:0/2 instruction, press F2 to branch to target *B*.
52. _____ Press F3, *Append Instruction*.
53. _____ Press F1, *Bit*.
54. _____ Press F1, *XIC*.
55. _____ Enter Bit Address > *B3:0/3*.
56. _____ Press *Enter* to accept the address.
57. _____ Press right arrow key so rung cursor will be on the right power rail.
58. _____ Press *Escape* to go to the branch selection screen.
59. _____ Press F1, *Branch*.
60. _____ Press F4, *Branch*.
61. _____ You are currently at target *C*. We need to branch around the XIO B3:0/3 instruction. Select target *B*, by pressing F2.
62. _____ Press *Append Instruction*, F3.
63. _____ Press F1, *Bit*.
64. _____ Press F2, *XIO*.
65. _____ Enter Bit Address > *B3:0/2*.
66. _____ Press *Enter* to accept address.
67. _____ Cursor to the right power rail.
68. _____ To program the OTE instruction, Press F1, *Bit*.
69. _____ Press F3, *OTE*.
70. _____ Enter Bit Address > *B3:0/3*.
71. _____ Enter to accept address.
72. _____ Press *Enter* or *Escape* to go to accept rung function selection screen.
73. _____ F10, *Accept Rung*.

RUNG TWO

74. _____ Press F4, *Insert Instruction*.
75. _____ Press F1, *Bit*.
76. _____ Press *XIC*.

77. _____ Enter Bit Address > *B3:0/3*.
78. _____ Press *Enter* to accept the address.
79. _____ Press F1, *Bit*.
80. _____ To enter the OTE instruction, press F3. This will be our real output instruction.
81. _____ Enter Bit Address > *O:2/1*.
82. _____ Press to accept address.
83. _____ Press *Enter* or *Escape* to go to the Accept Rung screen.
84. _____ Press F10, *Accept Rung*.
85. _____ Press *Escape* to remove the automatically appended rung.

DOWNLOAD AND TEST THE PROGRAM

1. _____ Save and go on-line to download your program to your PLC.
2. _____ Put the PLC in run mode.
3. _____ As your program runs verify correct operation after you understand how the program operates.
4. _____ Go off-line and add rung and instructions comments to rung one and two describing their operation.
5. _____ When completed with the program, save the program to your student floppy disk.

16

PROGRAMMING PRACTICE, LATCHING INSTRUCTIONS

For this exercise you will develop a ladder program using series, parallel, one-shot and the latch and unlatch instructions. Figure 16-1 on the next page illustrates the program you will develop.

1. _____ Copy the *Begin* file from your student floppy disk to your personal computers hard drive.
2. _____ Rename the *Begin* file *Ladder_6*.
3. _____ Escape to the program directory for processor.
4. _____ Select file by pressing F4, *Change File*.
5. _____ Use arrow keys to select *Ladder_6 file*.
6. _____ F9, *Save To File*.
7. _____ F1, *Offline Programming / documentation*.
8. _____ Press F8, *Monitor File*. A one-rung ladder with the word *End* on the rung will be displayed.
9. _____ Using your newly learned programming skills, complete the program as illustrated in Figure 16-1. Make sure you adjust the I/O addresses, if necessary, so they are correct for your particular PLC.
10. _____ When finished programming this ladder program, download and run the program to verify proper operation.
11. _____ Save program to your student floppy.

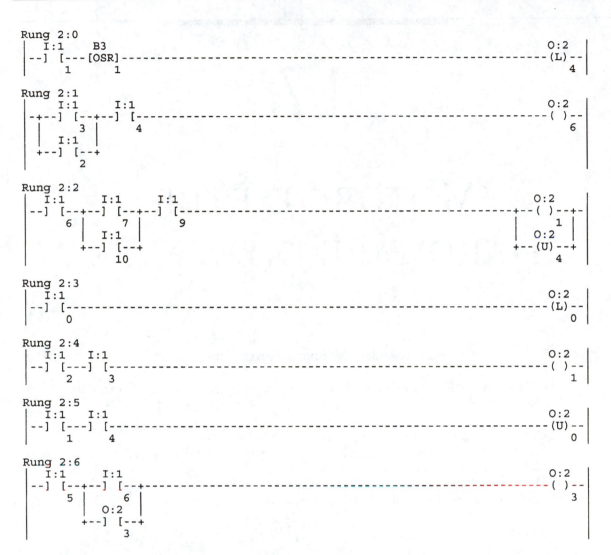

```
Rung 2:0
|  I:1      B3                                                        O:2    |
|--] [--- [OSR]---------------------------------------------------- (L) --  |
|   1       1                                                          4     |

Rung 2:1
|     I:1      I:1                                                     O:2    |
|-+--] [--+--] [--------------------------------------------------- ( ) --  |
| |    3  |    4                                                       6     |
| | I:1   |                                                                 |
| +--] [--+                                                                 |
|     2                                                                     |

Rung 2:2
|  I:1      I:1      I:1                                               O:2   +- 
|--] [--+--] [--+--] [------------------------------------------+-- ( ) --+- 
|     6 |     7 |    9                                          |     1   | 
|       | I:1   |                                              |   O:2   | 
|       +--] [--+                                              +-- (U) --+ 
|         10                                                        4       

Rung 2:3
|  I:1                                                                O:2    |
|--] [-------------------------------------------------------------- (L) -- |
|   0                                                                  0     |

Rung 2:4
|  I:1    I:1                                                         O:2    |
|--] [---] [-------------------------------------------------------- ( ) -- |
|   2      3                                                           1     |

Rung 2:5
|  I:1    I:1                                                         O:2    |
|--] [---] [-------------------------------------------------------- (U) -- |
|   1      4                                                           0     |

Rung 2:6
|  I:1      I:1                                                       O:2    |
|--] [--+--] [--+-------------------------------------------------- ( ) --  |
|     5 |     6 |                                                      3     |
|       | O:2   |                                                           |
|       +--] [--+                                                           |
|         3                                                                 |
```

Figure 16-1   Latching instruction program practice.

17

MOTOR STARTER
HARDWARE INTERFACE

This lab exercise will give you hands-on experience interfacing an Allen-Bradley modular SLC 500 PLC to control a motor starter. While completing this exercise, you will:

1. Convert a conventional relay ladder diagram to PLC format.
2. Create a PLC ladder and program from this converted ladder diagram.
3. Hook up and interface a start-stop station to the PLC.
4. Hook up a motor starter as an output device from your PLC.
5. Verify correct programming of a normally open start push button and a normally closed push button.

PREVIOUS KNOWLEDGE REQUIRED TO COMPLETE THIS LAB SUCCESSFULLY

Before attempting this lab exercise you should have a solid knowledge of wiring hardware relays, series and parallel hands-on wiring with 120 Volts AC line voltage, and have hooked up a similar motor starter with conventional hard wiring.

SAFETY FIRST!

1. _____ DANGER—120 VOLTS AC LINE VOLTAGE WILL BE PRESENT!
2. _____ Before beginning, have your instructor review and demonstrate this lab exercise so you know what to do, and what to expect as you proceed.
3. _____ Have your instructor review safety procedures for using your particular lab equipment.
4. _____ To ensure personal safety, have your instructor check your hookup before applying power!
5. _____ Examples and hookup exercises have been simplified for instructional purposes, and may not prove acceptable in actual industrial applications. Check and follow all applicable codes and ordinances when installing and working with electrical equipment.

TEXTBOOK REFERENCES

Study *Introduction to Programmable Logic Controllers* text, Chapter 12, "Understanding Relay Instructions" before attempting this hands-on lab. Review texts from your industrial electricity classes on motor starters and wiring.

INTRODUCTION

We will be using a typical motor starter, along with a start-stop push-button station, to interface to a PLC. Figure 17-1 illustrates the major parts of an Allen-Bradley Bulletin 509 motor starter.

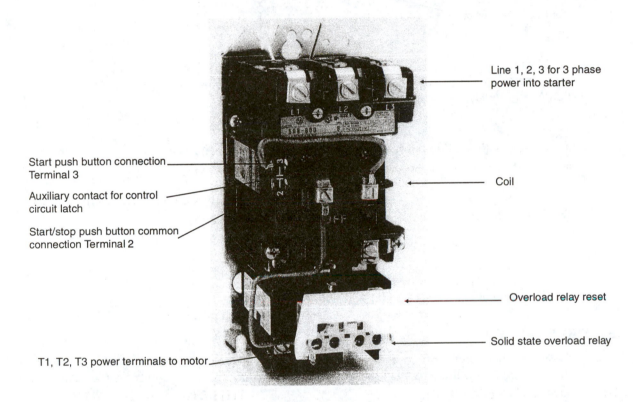

Line 1, 2, 3 for 3 phase power into starter

Start push button connection Terminal 3

Auxiliary contact for control circuit latch

Start/stop push button common connection Terminal 2

Coil

Overload relay reset

Solid state overload relay

T1, T2, T3 power terminals to motor

Figure 17-1 Major parts of an Allen-Bradley 509 motor starter. (Courtesy Rockwell Automation/ Allen-Bradley)

Conventional hookup of a motor starter consists of a push-button start-stop station hard-wired to the motor starter. Figure 17-2 (next page) illustrates typical conventional hard wiring of a motor starter. You need to be familiar with hard-wiring a motor starter before you complete this lab.

This lab exercise will guide you through the hookup and interface of a motor starter to an SLC 500 programmable logic controller.

ADDITIONAL MATERIALS NEEDED

To complete this lab exercise you will need:

1. Power, 120 volts AC, necessary tools and hookup wire.
2. Normally open start, normally closed stop push-button station.

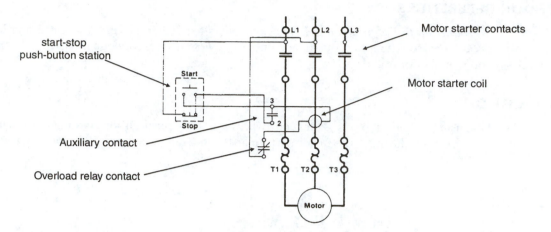

Figure 17-2 Conventional motor starter wiring diagram (compiled from Rockwell Automation/Allen-Bradley starter wiring data).

3. A motor starter similar to the Allen-Bradley Bulletin 509-BOD.
4. Allen-Bradley SLC 500 programmable logic controller.

To simplify wiring illustration for this exercise we will configure our PLC with the following two I/O modules:

1746 - IA8 input module.
1746 - OA8 output module.
IBM or compatible personal computer.
Rockwell Software's SLC 500 programming software.

GETTING STARTED

Be sure to check and verify your particular I/O modules and the correct wiring as you work through this exercise. We will start with the conventional motor starter ladder diagram illustrated in Figure 17-3.

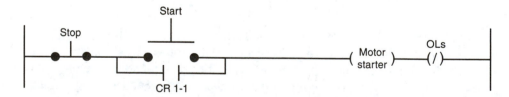

Figure 17-3 Conventional motor starter schematic.

RELAY LADDER CONVERSION TO PLC FORMAT

We need to work on converting our conventional motor starter wiring diagram to PLC format. This will enable us to develop a PLC ladder program and assist us in properly wiring inputs and outputs to our PLC I/O modules.

1. _____ The first step in relay schematic conversion is to determine the inputs and outputs.

   List all inputs:_____

   _____

   List all outputs:_____

   _____

2. _____ The second step is to allocate I/O reference numbers, or addresses. What are the addresses corresponding to each slot of your SLC 500 modular PLC? Figure 17-4 illustrates addresses for the 8-point input and output modules' screw terminals.

	Slot 0	Slot 1	Slot 2	Slot 3
		I:1.0	O:2/0	
		I:1/1	O:2/1	
		I:1/2	O:2/2	
Power supply	5/03 CPU	I:1/3	O:2/3	
		I:1/4	O:2/4	
		I:1/5	O:2/5	
		I:1/6	O:2/6	
		I:1/7	O:2/7	

Figure 17-4   Four-slot SLC 500 illustrating eight-point modules addressed in slots one and two.

3. _____ Assign your input addresses as in the table in Figure 17-5.

INPUT ADDRESS ASSIGNMENT	
List your inputs	Assigned addresses
Start normally open push button	I:0/0
Stop normally closed push button	I:0/1
Motor starter normally closed overloads	I:0/2
Motor starter normally open auxiliary	I:0/3

Figure 17-5   Input addresses have been assigned.

4. _____ Assign your output addresses as in Figure 17-6

OUTPUT ADDRESS ASSIGNMENT	
List your ouputs	Assigned addresses
Motor starter coil	O:2/0

Figure 17-6   The sole output address has been assigned.

5. _____ Draw the converted conventional relay ladder to PLC format.
6. _____ Figure 17-7 (next page) is a conceptional drawing of input wiring for the start/stop push button station to the PLC input module.

   Start-stop stations come in numerous varieties. Some start-stop stations come with only one set of normally open start contacts and one set of normally closed stop contacts. These contacts are built into a small enclosure called a start-stop push-button station. Other start-stop stations may have a removable contact block mounted on each push-button operator. These contact blocks may be single or double circuit. Depending on the particular start-stop station you have, you will have to wire it correctly.

   Wiring the push-button station in Figure 17-7: power comes in the left side of

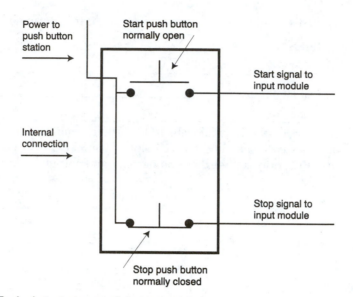

Figure 17-7 Typical start-stop push button station.

Figure 17-7 and is distributed to one side of each push button. The start push button is normally open and will normally send no signal to the input module screw terminal. Notice the stop push button is normally closed. Being normally closed, the Stop push button will continuously send an ON signal to the PLC input screw terminal.

7. _____ Figure 17-8 (next page) illustrates typical wiring for each of our inputs. There are four separate signals going into, or input into, the PLC input module. The 1746-IA8 module is an Allen-Bradley SLC 500 8-point, 120 volt AC input module. Power from line one goes to each hardware input device and then to the specified input address's screw terminal. The internally connected AC common screw terminals complete the circuit to line two. Notice that each input is completely independent as it sends its own signal into the module. Typically, inputs are separated when they are wired to a PLC. This separating of inputs is contrary to conventional hard wiring you may be accustomed to. Always refer to the manufacturer's wiring instructions before wiring your particular start-stop station.

8. _____ The only output from our PLC is to the motor starter's coil. This output will be wired as illustrated in Figure 17-9 (next page), assuming you are using the Allen-Bradley 1746-OA8 output module. To wire the module, power from line one goes into L-1 screw terminal. Think of the output module as a package containing a switch for each output address. When the program finds all rung inputs that are associated with our motor starter coil output as true, a logical *one* will be placed in the output status table, address O:2/0. When the processor updates the outputs, the ON signal in the output status table will be sent to the output module's internal solid state switch for address O:2/0. An ON signal to O:2/0 module address will close the solid state switch associated with that output. Power will flow to the starter coil, causing it to energize.

9. _____ Figure 17-10 (page 94) is a complete wiring diagram for our motor starter interface to the SLC 500 PLC. *This wiring diagram will only be correct if using the exact parts specified in the Additional Materials Needed section* (pp. 89–90). If using other than specified hardware, see your instructor for modified instructions. Notice that, to simplify Figure 17-10, only the top four I/O points are included for the I/O modules.

10. _____ In this lab exercise we are not going to hook up an actual motor to our motor starter. Your instructor can make motor installation an optional portion of the lab exercise. If a motor is to be hooked up, your instructor will supply needed instructions.

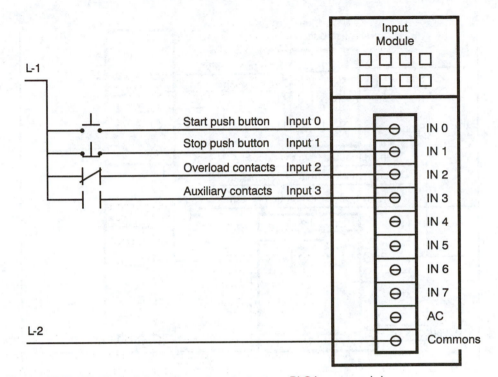

Figure 17-8 Typical wiring of a motor starter to a PLC input module.

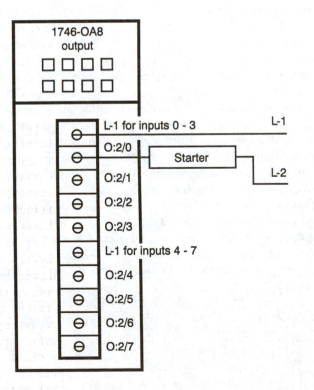

Figure 17-9 Typical wiring of a PLC output module to a motor starter coil.

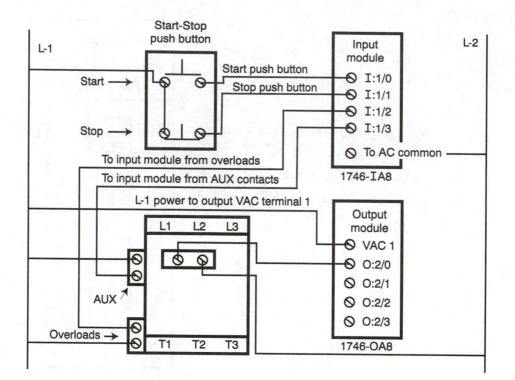

Figure 17-10 Typical motor starter interface to a PLC.

THE LAB

In the space provided, check off each step as completed

1. _____ Select all the parts needed to complete this lab.
2. _____ Place the processor and the correct I/O modules in the proper PLC chassis slots. Slot three will not be used. Place any available module in this slot.
3. _____ Do NOT apply power until your instructor has verified correct wiring!
4. _____ Connect a wire from the start push-button screw terminal to PLC input screw terminal I:1/0, as shown in Figure 17-10.
5. _____ Connect a wire from the stop push-button screw terminal to PLC input module screw terminal I:1/1 as illustrated in Figure 17-10.
6. _____ Connect a wire from the overload contact screw terminal to the input module screw terminal I:1/2, as illustrated in Figure 17-10.
7. _____ Connect a wire from output module screw terminal O:2/0 to one side of the motor starter coil.
8. _____ Connect a wire from auxiliary contact screw terminal to the input module screw terminal I:1/3 in Figure 17-10. This normally open contact, I:1/3, is the interlock around the Start push button.
9. _____ Connect the L-1 power wires to:
   Start-stop push-button station.
   Auxiliary contacts.
   Overload contacts.
   VAC 1 terminal on your 1746-OA8 output module.

10. _____ Connect the L-2 return wires to:
   Input module AC common screw terminal.
   One side of the motor starter coil.

11. _____ We do not need to connect a motor to verify proper PLC Interface. Consult with your instructor to see if connecting a motor will be added to this application.

12. _____ This should complete wiring of your control circuit. Do not apply power until completed and your instructor has checked your wiring.

13. _____ Develop your off line program on the personal computer using your SLC 500 development software. Remember to modify your I/O configuration if necessary.

14. _____ Connect the communications cable between the processor and your personal computer.

15. _____ Apply power to your SLC 500 PLC.

16. _____ Go on-line and download your program into the processor.

17. _____ Put the processor in run mode.

18. _____ IF your instructor has verified your control wiring is correct, apply power and test your inputs.

19. _____ When pressing the start button, does the motor starter energize?

20. _____ When you press the stop push button, does the motor starter disengage?

CONSIDERATIONS AS WE COMPLETE THIS APPLICATION

Now that you have successfully completed this lab exercise hooking up a motor starter and start-stop push-button station to a PLC, let us look at additional considerations for real-world applications.

Suppression of Electromagnetic Interference

Equipment surrounding a PLC may generate considerable amounts of *electromagnetic interference,* or *EMI.* Electromagnetic interference transients, or spikes, result from an inductive load's collapsing magnetic field when the device is switched off. Voltage levels from these spikes can be very high. EMI from noise-generating equipment such as relays, solenoids, and motor starters can cause intermittent problems in PLC operation, as well as damage output module circuitry. To help eliminate EMI transients, isolation transformers should be placed between a PLC and the AC voltage source. Surge suppression devices should also be used on all inductive devices operated by hard contact control devices. Hard contact devices include push buttons, selector switches, and relay contacts. Any inductive output device that is switched by hard contacts, including relay output PLC modules, need surge suppression. Even if the output control circuit is switched by a solid state output, and there is a hard contact switching device in the output circuit, surge suppression is necessary. Normally, a solid-state output module directly switching an inductive load may not require surge suppression. However, surge suppression can be added to the inductive load even though there are no hard contacts present in the output circuit, as there is still some degree protection for the solid-state device. Refer to your hardware manufacturer's instructions on selecting and installing surge suppression on your particular equipment.

Motor starters, typically NEMA size three or four and larger, draw too much current to be controlled by many PLC output modules. In this situation, a switching device needs to be placed between the PLC output module and the motor starter. This device, an *interposing relay*, can be easily switched by the PLC to control the larger sized motor starter.

Interposing Relay

When interfacing a PLC to any motor starter, or inductive load, the load's current draw must be considered in relation to the ratings of the output module. The point where the output module cannot handle the current is where an interposing relay is required in the output control circuit. When the control load is larger than the rating of the selected output module, a standard hard wired control relay is placed between the output module and the load. The output module switches the interposing relay, and the interposing relay switches the load. Care must be taken in selecting the interposing relay. The interposing relay must have inrush and sealed current values within the specifications of the output module. Most control relay contacts are rated for up to ten amps, so verify that the interposing relay contacts can handle the motor starter's coil load. Typically interposing relays are needed with size 3 or 4 and larger motor starters. Figure 17-11 lists the inrush and sealed coils' specifications for various starter sizes.

TYPICAL SEALED AND INRUSH VALUES FOR ALLEN-BRADLEY 3-POLE BULLETIN 509 STARTERS		
Starter size	Values listed in amps	
	Inrush value	Sealed value
0	1.67	.25
1	1.67	.25
2	2.09	.25
3	5.74	.39
4	10.62	.60
5	12.96	.835

Figure 17-11 Typical NEMA motor starters' current specifications. (Table data compiled from Rockwell Automation/Allen-Bradley data.)

DETERMINING THE LARGEST MOTOR STARTER OUR OUTPUT MODULE WILL SWITCH WITHOUT AN INTERPOSING RELAY

The 1746-OA8, a 100-240 VAC Triac output module, is rated at one amp continuous current per output point. The maximum surge current per output point for the 1746-OA8 is 10.0 amps for 25 milliseconds. From Figure 17-11, a size three NEMA starter will draw 5.74 amps inrush current, while a size four starter will draw 10.62 amps. Since this module is rated for 10.0 amps maximum surge current, a size four starter will need an interposing relay, even though continuous, or sealed current is well below maximum. Each output module chosen for an application should be checked for suitability before using the module.

QUESTIONS

1. Where would the motor leads have been connected to the motor starter if we had connected a motor to our starter?_____

2. If we had hooked up a motor to our motor starter, where does the power to operate the motor connect?_____

3. What is typically done to obtain a lower control circuit voltage level than the usual 480-volt, 3-phase motor voltage?_____

_____

4. Regarding the PLC program rung for controlling the motor starter, explain the extra set of contacts on the parallel branch around the start button?_____

_____

5. Why is the stop push button the first instruction on our PLC ladder rung?_____

6. The normally closed, stop push button is programmed as normally open. Would it not make sense to have the input instruction reflect the actual state of the push button? Explain._____

7. What would have happened if you had programmed your stop push-button as a normally closed instruction? Explain your answer. Did you try it?_____

8. Why is surge suppression needed on an inductive load?_____

9. Discuss any problems that may be encountered interfacing larger motor starters to PLC's.__

10. The overload contacts send a (normally open or normally closed?) signal to the PLC input module under normal running conditions.

11. When finished developing your program off line, you go on line. Explain what this means.__

12. Fill in the table in Figure 17-12 as to whether surge suppression is necessary under the described conditions. (Surge suppression can be added to inductive loads directly switched by solid state output modules to receive a degree of protection against transient spikes.)

IS SURGE SUPPRESSION NECESSARY?			
Output Module Type	Hard Contact Hardware in Output Circuit?	Field Device	Suppression Needed?
Solid state	No	Motor starter,	
Solid state	Yes, Selector switch	Transformer pilot light	
Relay output	Yes, Push button	Transformer pilot light	
Solid state	No	Interposing relay	
Relay output	No	Interposing relay	
Solid state	Yes, Selector switch	Solenoid	
Solid state	Yes, Push button	Pilot light	
Relay output	No	Solenoid	

Figure 17-12

SUMMARY

This chapter provided the opportunity to get hands-on experience hooking a motor starter to a PLC. You had the opportunity to prove that the normally closed, stop push button must be programmed as a normally open PLC input instruction. Likewise, the normally closed overload contacts must also be programmed as normally open PLC instructions.

Interposing relays are used when an output module cannot switch the current level required to energize and de-energize the motor starter coil. Remember, whenever hard contacts are controlling an inductive load, surge suppression is necessary. Always refer to your hardware documentation when selecting surge suppression for a specific application.

18

UNDERSTANDING RELAY INSTRUCTIONS

This exercise will give you practice selecting the proper instruction, or practice creating ladder rungs for a specific specification. Complete each question regarding proper instruction selection or ladder rung creation.

1. If input bit I:7 is normally a *one* and the rung needs to be true when the input status table bit is a *zero,* what instruction should be programmed on the PLC ladder rung?

2. Input bit I:2 is a *one,* input bit I:8 is a *one,* the rung to be true when both input status table bits are *ones.*
   A. Draw this PLC ladder rung. Use O:2/0 as your output.
   B. What instructions will be programmed on your PLC ladder rung?_____
   _____
   C. What happens if the I:8 input status bit goes false?_____
   _____

3. If input bit I:2 is a one, input bit I:8 is a one; develop a rung that will be true when either of these input status table bits is a one.
   A. Draw the PLC ladder rung. Use output O:2/1 as the output.
   B. What instructions were programmed on the PLC ladder rung?_____
   _____
   C. What happens if I:8 status bit goes false?_____
   _____

4. We have a piece of machinery with a table where the part being manufactured is held. The table moves, allowing work to be done by the machine on the part. Currently the table is in position holding a normally open limit switch closed. What PLC instruction should be programmed on the ladder rung so the instruction would be true when the table moved away from

   the limit switch?_____

5. We have an inductive proximity switch looking at a drill bit. An inductive proximity sensor is used to sense the presence of metal. The proximity switch is a normally open style, so that it is closed when the drill bit is seen by the switch. We want the PLC to send an alarm to the operator if the drill bit breaks, or if the cable from the drill bit sensor were to become damaged and no signal was being sent. What instruction would be programmed so the rung becomes true in an alarm condition? (When an alarm condition is detected, the machine shuts down and the alarm bell sounds.)_____

6. A robot is busy assembling our product. Around the cell are clear plastic doors that are to stay closed unless a maintenance person needs to go in and repair the robot. Each door has a switch connected to a PLC input. If any door is accidentally opened, the robot is to stop all movement. If a forklift were to hit the cell and cut the cable, the robot is also to stop all movement. In case of switch failure or damage, the robot must be fail-safe.

A. Should a normally open or a normally closed switch be installed on each of the four doors?

_____

B. What instruction should be programed on your PLC ladder rung?

_____

C. Throughly explain your choices._____

_____

7. Figure 12-2 in the text refers to the incorrect conversion of a conventional, start-stop schematic to PLC control. Explain in detail why programming a PLC ladder in this manner will not operate correctly._____

_____

8. When developing a typical start-stop rung of PLC logic, why should the stop push button be programmed as the first instruction on the rung?_____

_____

9. Explain what is meant when OSHA requires that a stop push button be fail-safe._____

_____

10. When converting a conventional schematic to PLC control what is the first step?_____

_____

11. Why do we separate inputs so each input provides a separate input signal as an input?_____

_____

12. Illustrate a typical 8-point input module and how it would be wired in a circuit with two limit switches in parallel, and then two limit switches in series. Include input addresses in your drawing.

13. A normally open, held closed limit switch is wired into I:1/5. What status bit would be found, and in what position of the 16-bit word representing the input module?_____

_____

19

PLC SYSTEM DOCUMENTATION

The ladder diagram printout, Figure 19-1 on the next page, has parts identified with numbers. For the following exercise, match those numbers with the term that best identifies the portion of the ladder diagram identified.

Number 1 is: _____   A. Left power rail

Number 2 is: _____   B. Right power rail

Number 3 is: _____   C. Examine if open contact

Number 4 is: _____   D. Examine if closed contact

Number 5 is: _____   E. Output energize instruction

Number 6 is: _____   F. Rung comment

Number 7 is: _____   G. Instruction comment

Number 8 is: _____   H. Bit address

Number 9 is: _____   I. File number

Number 10 is: _____   J. Rung Number

Number 11 is: _____   K. Input module identification

Number 12 is: _____   L. Output module identification

Number 13 is: _____   M. End of program rung

Number 14 is: _____   N. Input instruction address

Number 15 is: _____   O. Output instruction address

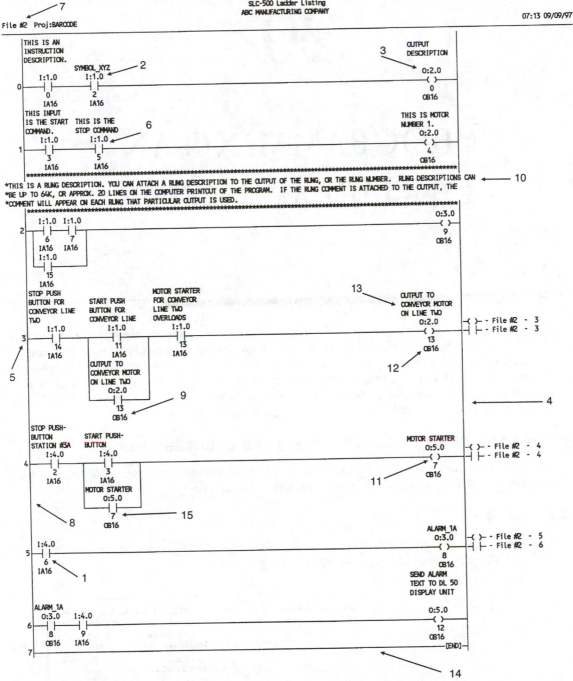

Figure 19-1 Ladder diagram print out from Rockwell Software's PLC 500 software.

20

PROGRAMMING AN
ON-DELAY TIMER

This lab exercise will give you practice developing a ladder program containing an on-delay timer. When finished developing the ladder logic, you will go on-line, download and rung the program.

Use the on-delay timer instruction to program a time delay before instruction becomes true. When you want an action to begin a specified time after the input becomes true. As an example, a certain step in the manufacturing process is to begin 30 seconds after a signal is received from a limit switch. The 30-second delay is the on-delay timer's preset value.

KNOWLEDGE REQUIRED TO COMPLETE THIS LAB SUCCESSFULLY

You should have completed study of Chapter 14 of *Introduction to Programmable Logic Controllers*. Refer to that chapter for information on timers and counters.

THE LAB, PART I

Part one of this timer programming lab exercise will step you through programming, downloading and running the timer ladder logic illustrated in Figure 20-1.

```
Rung 2:0
 | I:1                                                      +TON---------------+
 |--] [--------------------------------------------------+TIMER ON DELAY    +-(EN)-
 |   0                                                    |Timer        T4:0+-(DN)
 |                                                        |Time Base     1.0|
 |                                                        |Preset         30|
 |                                                        |Accum           0|
 |                                                        +-----------------+

Rung 2:1
Timer four's done bit will energize output O:2/1 after the time delay of thirty
seconds.
 | T4:0                                                                    O:2
 |--] [------------------------------------------------------------------( )--
 |   DN                                                                     1
```

Figure 20-1   Part one timer programming ladder.

We will begin developing our PLC ladder program using our SLC 500 programming software.

1. _____ Load a copy of the *Begin* program from your student floppy disk. If you modified your PLC I/O configuration for the last exercise, return your hardware to the default configuration. Press F 3, *Offline Programming and Documentation*.
2. _____ Rename to *Timer_1*.
3. _____ Edit Your *Timer_1* program.

Creating Rung Zero

4. _____ Press F4, *Insert Rung*.
5. _____ Press F4, *Insert Instruction*.
6. _____ Press F1, *Bit Instruction*.
7. _____ Press F1, *XIC Instruction*.
8. _____ Type in address *I:1/0*.
9. _____ Press *Enter*.
10. _____ Press F2, *Timer / Counter Instructions*.
11. _____ Press F1, *TON (On-Delay Timer)*.
12. _____ Enter timer address> *T4:0*.
13. _____ Press *Enter*.
14. _____ Fill in Time Base with *1.0* (seconds).
15. _____ Press *Enter*.
16. _____ Fill in Preset Value with *30* (seconds).
17. _____ Press *Enter*.
18. _____ Fill in the accumulated value with *0*.
19. _____ Press *Enter*. This rung is complete.
20. _____ Press the *Escape* key.
21. _____ Accept the rung by pressing F10.

Creating Rung One

22. _____ Press F3, *Append Rung*.
23. _____ Press F4, *Insert instruction*.
24. _____ Press F1, *Bit Instruction*.
25. _____ Press F1, *XIC*.
26. _____ Enter Bit Address: *T4:0/DN*.

Entering the Rung's Output Instruction

27. _____ Press F1, *Bit*.
28. _____ OTE (output enable).
29. _____ Enter Bit Address: *O:2/1*.
30. _____ Press *Enter*.
31. _____ Press *Escape*.
32. _____ Press F10 to accept rung.
33. _____ Press *Escape* to remove automatically appended rung.
34. _____ Cursor back to the T4:0/DN instruction.

Adding a Rung Comment

Next we will add a rung comment—but before we do, let's check your display configuration to verify that documentation will be displayed on the computer screen.

A. Press *Escape*.

B. F2, *Config Display.*

C. Toggle F7 so it displays *Suppress Rung Comment* if you want to see the comment text on your screen.

D. Toggle F7 so it displays *Display Rung Comment* if you do not want to see the comment text on your screen.

35. _____ Press *Escape.*
36. _____ Press F5, *Document.*
37. _____ Press F1, *Rung Comment.*
38. _____ Type in the following: *Timer four's done bit will energize output O:2/1 after the time delay of thirty seconds.*
39. _____ Press F8, *Accept / Exit.*
40. _____ Press F10, *Save Documentation.*
41. _____ *Escape.*
42. _____ Press F10, *Edit.*
43. _____ Press *Save / Go On* line to download.
44. _____ Put processor in Run mode.
45. _____ Energize input I:1/0.
46. _____ Explain what happens.
47. _____ De-energize the timer's input, I:1/0. Explain what happens.

48. How is an ON-Delay timer reset?_____

49. What is the purpose of the preset parameter of the timer instruction?_____

_____

50. What is the purpose of the accumulated value of the timer instruction?_____

_____

51. What is the timer's time base?_____

52. The programming parameter Timer T4:0 serves what purpose?_____

_____

53. This completes programming an On-Delay timer.
54. If you will be continuing with lab work, for this session skip to step one for the next exercise.
55. If you are finished for today, save this program to your floppy disk Name your program *Timer_1.*
56. Exit the software and shut down your computer.

ON-DELAY PROGRAMMING LAB EXERCISE, PART II

This programming exercise is going to edit the existing program *Timer_1*. You will insert two new rungs between current rungs 2:0 and 2:1. The completed program is illustrated in Figure 20-2 on page 106.

Begin with the previous ladder diagram.

1. _____ Starting at the monitor File Screen.
2. _____ Press F10, to edit your ladder program.
3. _____ Move the cursor to rung 2:0.

4. _____ Press F3, *Append Rung*. A blank rung will be added between the two original rungs. All affected rungs will be automatically renumbered.

        Note: An alternative method of inserting a rung here would be to put the cursor on rung 2:1 and press F4, *Insert Rung*.

5. _____ Press F4, *Insert Instruction*.
6. _____ Press F1, *Bit*.
7. _____ Press F1, *XIC*.
8. _____ Enter Bit Address> T4:0/EN.

Add Output Instruction To Rung

9. _____ Press F1, *Bit*.
10. _____ Press F3, *OTE*.
11. _____ Enter Bit Address> O:2/2.
12. _____ *Escape*.
13. _____ F10, *Accept Rung*.

Add Rung Documentation

14. _____ Press *Escape*.
15. _____ Press F5, *Document*.
16. _____ Put cursor on the T4:0/EN instruction.
17. _____ Press F2, *Instruction Comment*.
18. _____ Type in the following text: *Timer 4:0 Enabled*.
19. _____ Press F8, *Accept / Exit*.
20. _____ Press F1, *Rung Comment*.
21. _____ Type in the rung comment as shown in figure 20-2.
22. _____ Press *Accept / Exit*.
23. _____ Press F10, *Save Documentation*.

Add the Next Rung

24. _____ Move cursor to left power rail of newly created rung.
25. _____ Press *Escape*.
26. _____ Press F10, *Edit*.
27. _____ Press F4, *Insert Rung*.
28. _____ Press F4, *Insert Instruction*.
29. _____ Press F1, *Bit*.
30. _____ Press F1, *XIC Instruction*.
31. _____ Enter Bit Address > T4:0/TT (This is the timer timing bit.)
32. _____ Press *Enter*.

The Output Instruction

33. _____ Press F1, *Bit Instruction*.
34. _____ Press F3, *OTE Instruction*.
35. _____ Enter Bit Address> O:2/3.
36. _____ Press *Enter*.
37. _____ Press *Escape*.
38. _____ Press F10 to accept the rung.

```
Rung 2:0
    I:1                                                          +TON---------------+
 --] [---------------------------------------------------------+TIMER ON DELAY    +- (EN) -
       0                                                        |Timer           T4:0+- (DN)
                                                                |Time Base        1.0|
                                                                |Preset            30|
                                                                |Accum              0|
                                                                +-----------------+

Rung 2:1
The timer timing bit is set when rung conditions are true and the accumulated
value is less than the preset value. The timer timing bit remains set until
rung conditions go false, or the done bit is set.
    Timer
    timing
    input bit
        T4:0                                                                    O:2
 ----] [-----------------------------------------------------------------------( )--
        TT                                                                       3

Rung 2:2
The timer enable bit is set when the timer instruction's rung is true.
    Timer 4:0
    enabled.
        T4:0                                                                    O:2
 ----] [-----------------------------------------------------------------------( )--
        EN                                                                       2

Rung 2:3
Timer four's done bit will energize output O:2/1 after the time delay of
thirty seconds.
    T4:0                                                                        O:2
 --] [-------------------------------------------------------------------------( )--
        DN                                                                       1
```

Figure 20-2 On-delay programming exercise, part two.

Documentation

39. _____ Press *Escape*.
40. _____ Press F5, *Documentation*.
41. _____ Cursor to the T4:0/TT instruction.
42. _____ Press F2, *Instruction Comment*.
43. _____ Type in the instruction comment shown for this instruction from Figure 20-2.
44. _____ When completed typing instruction comment, press F8, *Accept /Exit*.
45. _____ Press F1, to add a rung comment.
46. _____ Type in the rung comment as shown in Figure 20-2.
47. _____ When completed typing in the rung comment press F8 to accept and exit.
48. _____ Press F10 to save documentation.
49. _____ Press *Escape*.
50. _____ Press F10 to edit ladder.
51. _____ Press F1, *Save and go on line*.
52. _____ Put the PLC in run mode.
53. _____ Close I:1/0.

As Your Program Runs, Answer the Following Questions

1. Explain the operation of an on-delay timer._____
_____

2. Define each part of the timer's address for T4:0.

    T = _____

    4 = _____

    : = _____

    0 = _____

3. Define each parameter of the timer instruction:

    Timer = _____

    Time base = _____

    Preset = _____

    Accumulated Value = _____

4. When power is applied to input I: 1/0, explain the operation of the timer._____
_____

5. When the timer's preset value and the accumulated value are equal, what allows us to control
    other logic on our ladder program?_____
_____

6. What happens if a TON instruction looses power?_____
_____

7. How is the timer in this exercise reset?_____

8. Explain the operation of the timer timing bit._____
_____

9. Explain the operation of the timer done bit._____
_____

10. Explain the operation of the timer enable bit._____
_____

11. _____ Save your program to floppy disk as *Timer_2*, for future editing.
12. _____ Ask your instructor if you need to print out your ladder for handing in.
13. _____ If finished with this lab session, shut down your computer.

ON-DELAY PROGRAMMING LAB EXERCISE, PART III

For this exercise you will edit program file *Timer_2* and make it into a free-running timer.

Timer T4:0, from Part II of the on-delay programming exercise, times until the preset is reached and the done bit comes on. The done bit turns on output O:2/1 and the output stays on until we reset the timer by de-energizing input I 1/0. Whenever power is lost to a timer input, the timer resets itself to zero, We can use this principle to make our previous timer a free-running timer.

A free-running timer will time up to its preset value, set the done bit, reset itself, and start the timing cycle over. The timer will output a pulse every 30 seconds. We will edit the ladder and T4:0 in program file *Timer_2* to operate as a free-running timer in this exercise. A pictorial representation of the timer done bit's behavior is illustrated in Figure 20-3.

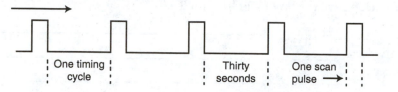

Figure 20-3   Timing diagram for free-running timer for this exercise.

Every 30 seconds (the preset interval in this particular application) is the time between timer cycles. This pulse can be used with other logic in our ladder program. This exercise will edit program file *Timer_2* and make it into a free running timer. The completed program is illustrated in Figure 20-4 on the next page. An XIO T4:0/DN instruction will be placed in series with the input instruction and the timer instruction from program file *Timer_2*.

From the file options screen, copy *Timer_2* and name the copied file *Timer_3*. Load *Timer_3* program file.

1. _____ From the monitor file screen, press F10
2. _____ Edit rung zero as follows:
4. _____ Move cursor to the I:1/0 instruction.
5. _____ Press F5, *Modify Rung*.
6. _____ Press F3, *Append Instruction*.
7. _____ Press F1, *Bit*.
8. _____ Press F2, *XIO*.
9. _____ Enter bit address T4:0/DN.
10. _____ Press *Enter*.
11. _____ Press *Escape*.
12. _____ Press F10, *Accept Rung*.

Add Documentation

13. _____ Press *Escape*.
14. _____ Press F5, *Document*.
15. _____ Press F1, *Rung Comment*.
16. _____ Add rung comment as illustrated in Figure 20-4.
17. _____ Press F8, *Accept / Exit*.
18. _____ Press F10, *Save documentation*.
19. _____ Press *Escape*.
20. _____ Press F10, *Edit*.
21. _____ Press F1, *Save and go online*.
22. _____ Go to run mode.
23. _____ When finished testing your program, save the program to your floppy disk.

As you evaluate program execution, answer the following questions:

1. When running the program, explain how the output O:2/1 LED behaves. Explain why._____

```
Rung 2:0
The T4:0/DN XIO instruction makes this a free running timer.  As a free
running timer, the timer done bit will be energized every thirty seconds. The
done bit will energize for one processor scan. Thirty seconds is the preset
value. Each time the preset is equal to the accumulated value, the done bit
will be made true. Being true, the done bit will remove power from the TON
instruction for one scan, causing the instruction to reset to zero.
    I:1   T4:0                                          +TON---------------+    |
--] [---]/[-----------------------------------------+TIMER ON DELAY     +-(EN)-
     0     DN                                        |Timer         T4:0+-(DN)
                                                     |Time Base      1.0|
                                                     |Preset          30|
                                                     |Accum            0|
                                                     +-----------------+    |

Rung 2:1
The timer timing bit is set when rung conditions are true and the accumulated
value is less than the preset value. The timer timing bit remains set until
rung conditions go false, or the done bit is set.
    Timer                                                                  |
    timing                                                                 |
    input bit.                                                             |
       T4:0                                                          O:2   |
----] [------------------------------------------------------------( )--
       TT                                                            3  |

Rung 2:2
The timer enable bit is set when the timer instruction's rung is true.
    Timer 4:0                                                              |
    enabled.                                                               |
       T4:0                                                          O:2   |
----] [------------------------------------------------------------( )--
       EN                                                            2  |

Rung 2:3
Timer four's done bit will energize output O:2/1 after the time delay of
thirty seconds.
    T4:0                                                               O:2   |
--] [--------------------------------------------------------------( )--
    DN                                                              1  |
```

Figure 20-4   *Timer_3* ladder program.

2. Explain the sequence of events beginning with input I:1/0 becoming true, and up to the point

of timer reset and resumption of another timing cycle._____

_____

_____

21

PROGRAMMING AN OFF-DELAY TIMER

This programming exercise you will give you practice developing a ladder program with an off-delay timer. This will cause a time delay to begin after rung inputs go false. As an example, if a cooling fan is to run all the time the motor is running and for five minutes after motor is turned off, use a five minute off-delay timer. The five-minute timing cycle begins when the motor is turned off.

THE LAB

1. _____ Load the begin program from your student floppy disk to your personal computer's hard drive.
2. _____ Use the file options to rename the begin file to *OFF DELAY*.
3. _____ Load the *OFF DELAY* processor file so you can edit the file. The ladder program for this exercise is illustrated in Figure 21-1 on the next page.

ENTERING RUNG ZERO

4. _____ When you are at the program edit screen, press F4, *Insert rung*.
5. _____ Press F4, *Insert Instruction*.
6. _____ Press F1, *Bit*.
7. _____ Press F1, *XIC*.
8. _____ Enter Bit Address >*I:1/0*.
9. _____ Press *Enter* to accept the address.
10. _____ Press F2 to select the timer and counter instructions selections.
11. _____ Press F2 to select the Timer Off Delay (TOF) timer.
12. _____ Enter Timer Address > *T4:0*.
13. _____ Press *Enter*.
14. _____ Enter Time Base > *1.0*. This gives a one second time base.
15. _____ Press Enter to select this parameter.

```
Rung 2:0
The timer off delay instruction (TOF) is used to either turn an output on or
off after its rung has been false for the preset time interval. The TOF
instruction begins incrementing toward the preset when the rung transitions
from true to false. As long as rung conditions remain false, the timer will
increment toward its programmed accumulated value. The accumulated value will
be cleared to zero anytime the rung transitions from false to true.
|    I:1                                                +TOF---------------+
--] [-------------------------------------------------+TIMER OFF DELAY  +-(EN)-
|    0                                                 |Timer       T4:0+-(DN)
|                                                      |Time Base     1.0|
|                                                      |Preset         30|
|                                                      |Accum           0|
|                                                      +-----------------+
```

```
Rung 2:1
The done bit is true when rung conditions transition from false to true. The
done bit continues to be true  when the rung goes false and the accumulated
value is less than the preset value. This instruction will keep the done bit
energized for 30 seconds, the preset value, after input I:1/0 goes false.
|    T4:0                                                                O:2
--] [--------------------------------------------------------------------( )--
|    DN                                                                   1
```

```
Rung 2:2
The timer timing bit (TT) is only true when the timer's accumulated value is
accumulating toward the preset. The only time the timer times is when the rung
goes false, and the off-delay portion of the cycle is active.
|    T4:0                                                                O:2
--] [--------------------------------------------------------------------( )--
|    TT                                                                   2
```

```
Rung 2:3
The enable bit is true whenever the rung's input is true.  When I:1/0 is true,
the enable bit will be true.  When input I:1/0 is false, eventhough the timer
may be timing during the off-delay portion of the cycle, the EN bit will be
false.
|    T4:0                                                                O:2
--] [--------------------------------------------------------------------( )--
|    EN                                                                   3
```

Figure 21-1   Off-delay timer ladder program for this exercise.

16. _____ Enter Preset Value > *30*. This gives us a thirty-second preset value.
17. _____ Press *Enter* to select this parameter.
18. _____ Enter Accumulated Value > *0*. This gives a starting accumulated value of zero.
19. _____ Press *Enter* to select this parameter.
20. _____ Press *Enter* or *Escape* to go to the Accept Rung function.
21. _____ Press F10 to *Accept Rung*.

ENTERING RUNG ONE

22. _____ Press F3, *Append Rung*.
23. _____ Press F4, *Insert Instruction*.
24. _____ Press F1, *Bit*.
25. _____ Press F1, *XIC*.
26. _____ Enter Bit Address > *T4:0/DN*. This is timer four's done bit.
27. _____ Press F1, *Bit*. We will select the rung's output instruction next.
28. _____ Press F3 to select the OTE instruction.
29. _____ Enter Bit Address > *O:2/1*.

30. _____ Press *Enter* to accept address.
31. _____ Press *Enter* or *Escape* to go to the Accept Rung function.
32. _____ Press F10 to *Accept Rung*.

ENTERING RUNG TWO

33. _____ Press F4 to insert an instruction on the appended rung.
34. _____ Press F1, *Bit*.
35. _____ Press F1, *XIC*.
36. _____ Enter Bit Address > *T4:0/TT*. This is timer four's timer timing bit.
37. _____ Press *Enter* to accept this address.
38. _____ Press F1, *Bit* to select the output instruction.
39. _____ Press F3, *OTE*.
40. _____ Enter Bit Address > *O:2/2*.
41. _____ Press *Enter* to accept the address.
42. _____ Press *Enter* or *Escape* to go to the Accept Rung function.
43. _____ Press F10 to *Accept Rung*.

ENTERING RUNG THREE

44. _____ The software should have automatically appended a blank rung.
45. _____ Press F4, *Insert Instruction*.
46. _____ Press F1, *Bit*.
47. _____ Press *XIC*.
48. _____ Enter Bit Address > *T4:0/EN*. This is the timer's enable bit.
49. _____ Press *Enter* to accept this address.
50. _____ Press F1, *Bit* to select the rung's output instruction.
51. _____ Press F3, *OTE*.
52. _____ Enter Bit Address > *O:2/3*.
53. _____ Press *Enter* to accept the address.
54. _____ Press *Enter* or *Escape* to go to the Accept Rung function.
55. _____ Press F10 to *Accept Rung*.
56. _____ Press *Escape* to delete the automatically appended rung.

ADDING DOCUMENTATION

1. _____ Use the arrow keys to move cursor to rung zero.
2. _____ Press *Escape* to go to the documentation function key selection.
3. _____ Press F5, *Document*.
4. _____ Press F1, *Rung Comment*.
5. _____ Add documentation as illustrated in Figure 21-1 (previous page) for this rung.
6. _____ When completed, press F8, *Accept / Exit*.
7. _____ Use arrow keys to cursor to rung one.
8. _____ Press F1, *Rung Comment*.
9. _____ Add documentation as illustrated in Figure 21-1 for this rung.
10. _____ When completed, press F8, *Accept / Exit*.
11. _____ Use arrow keys to cursor to rung two.
12. _____ Press F1, *Rung Comment*.
13. _____ Add documentation as illustrated for this rung.
14. _____ When completed, press F8, *Accept / Exit*.

15. _____ Use arrow keys to cursor to rung three.
16. _____ Press F1, *Rung Comment*.
17. _____ Add documentation as illustrated for this rung.
18. _____ When completed, press F8, *Accept / Exit*.
19. _____ Press F10, *Save Documentation*.
20. _____ Press *Escape*.
21. _____ Press F10, *Edit*.
22. _____ Press F1, *Save / Go On Line*.
23. _____ Answer F8, *Yes*, to the message, "Change Processor to Run Mode?"
24. _____ As you experiment with the timer operation, fill in the table in Figure 21-2.

I:1/ Input	DN —] [—	TT —] [—	EN —] [—
ON			
OFF[1]			
OFF[2]			

[1] Timer starts timing.
[2] Timer preset equals accumulated value.

Figure 21-2   TOF status bits operation.

EDITING YOUR PROGRAM

Now that you have experimented with the TOF timer program, let's go off-line and do some editing.

1. _____ Press F10, *Edit*.
2. _____ Let's edit the hard drive program, Press F3, *Off Line Disk*.
3. _____ Cursor to rung one. This is the rung with the done bit.
4. _____ Press F3, *Append Rung* to add a new rung between current rung one and rung two. We are going to add one new rung of logic. This is illustrated in Figure 21-3 on the next page.
5. _____ Press F4, *Insert instruction*.
6. _____ Press F1, *Bit*.
7. _____ Press F2, *XIO*.
8. _____ Enter Bit Address > *T4:0/DN*.
9. _____ Press *Enter* to accept address.
10. _____ Press F1, *Bit*. We will now enter the output instruction.
11. _____ Press F3, *OTE*.
12. _____ Enter Bit Address > *O:2/4*.
13. _____ Press *Enter* to accept the address.
14. _____ This is the extent of our editing. Press *Enter* or *Escape* to go to the Accept Rung Function selection.
15. _____ Press F10, *Accept Rung*.
16. _____ Press *Escape* to remove the automatically appended rung.
17. _____ Press F1, *Save / Go On Line*.
18. _____ Answer F8, *Yes* to the message, "File already exits, overwrite file?" This message tells you that there is a file with the same name already in the processor. "Do you want to write over the current processor's file?" The file on your personal computer's hard drive is the edited version of the program. The program in the PLC is the old, unedited program. We do want to replace, or overwrite, the older processor program.

Rung 2:0
The off-delay timer instruction (TOF) is used to either turn output on or off
after its rung has been false for the preset time interval. The TOF
instruction begins incrementing toward the preset when the rung transitions
from true to false. As long as rung conditions remain false, the timer will
increment toward its programmed accumulated value. The accumulated value will
be cleared to zero anytime the rung transitions from false to true.

```
  I:1                                                        +TOF-------------+
--] [----------------------------------------------------+TIMER OFF DELAY   +-(EN)-
     0                                                    |Timer          T4:0+-(DN)
                                                          |Time Base       1.0|
                                                          |Preset           30|
                                                          |Accum             0|
                                                          +-----------------+
```

Rung 2:1
The done bit is true when rung conditions transition from false to true. The
done bit continues to be true when the rung goes false and the accumulated
value is less than the preset value. This instruction will keep the done bit
energized for 30 seconds, the preset value, after I:1/0 goes false.

```
  T4:0                                                                      O:2
--] [---------------------------------------------------------------------( )--
    DN                                                                       1
```

Rung 2:2
```
  T4:0                                                                      O:2
--]/[---------------------------------------------------------------------( )--
    DN                                                                       4
```

Rung 2:3
The timer timing bit is only true when the timer's accumulated value is
accumulating toward the preset. The only time the timer times is when the rung
goes false, and the off-delay portion of the cycle is active.

```
  T4:0                                                                      O:2
--] [---------------------------------------------------------------------( )--
    TT                                                                       2
```

Rung 2:4
The enable bit is true whenever the rung's input is true. When input I:0/1 is
true, the enable bit will be true. When input I:0/1 is false, even though the
timer may be timing during the off-delay portion of the cycle, the EN bit will
be false.

```
  T4:0                                                                      O:2
--] [---------------------------------------------------------------------( )--
    EN
```

Figure 21-3 Edited ladder program for this exercise.

19. _____ Answer *Yes,* F8, to the message: "Change processor mode to program?" The processor is currently in run mode. You have to change the PLC processor to program mode before we can download a new program.

20. _____ After the new program has been downloaded, the message: "Program successfully restored to processor. Change processor to run mode?" Press F8, *Yes.*

21. _____ As you experiment with the TOF program, fill in the table in Figure 21-4 (next page).

22. _____ Explain how the XIO done bit operates._____

_____

23. _____ Draw a timing diagram below illustrating both the XIC done bit and the XIO done bit.

24. _____ Go off-line and edit your personal computer's hard drive's ladder program. Add documentation to your newly added rung. Add a rung comment explaining how the done bit behaves.

25. _____ When completed, save the program to your student floppy disk.

I:1/ Input	O:2/1 DN —I I—	O:2/4 DN —I/I—	O:2/2 TT —I I—	O:2/3 EN —I I—
OFF[1]				
ON[2]				
OFF[3]				
OFF[4]				

[1] Input I:1/1 is false, has not been energized yet.
[2] Input I:1/1 is true, timer not timing.
[3] Input I:1/1 is false, timer timing.
[4] Input I:1/1 is false, accumulated value equal to preset.

Figure 21-4 TOF timer and associated status bits data.

22

PROGRAMMING A RETENTIVE TIMER

This programming exercise will edit a previous ladder program, *Timer_3,* and add a retentive timer instruction in place of the TON instruction. Figure 22-1 (next page) illustrates the completed program.

A retentive timer is used to retain accumulated value through a power loss, processor mode change, or rung state going from true to false. As an example, a retentive timer would be used to track the running time of a motor for maintenance purposes. Each time the motor is turned off the timer will remember the elapsed time; next time the motor is turned on, recorded time will increase from there. When you want to reset this timer, use a reset instruction.

THE LAB

1. _____ Copy the program *Timer_3* from your floppy disk.
2. _____ From the monitor file screen select *Edit,* F10.
3. _____ Move cursor to Rung 3.
4. _____ Cursor to XIO, T4:0 DN instruction.
5. _____ Press F5, *Modify Rung.*
6. _____ Press F6, *Delete instruction.*
7. _____ The T4:0 / DN instruction is deleted. If you delete an instruction in error you can press F7, to undelete the instruction.
8. _____ Try it yourself. Notice caution message, "Removing data references leaves data/forces in their last state."
9. _____ If you have undeleted the instruction, delete it. Once the instruction is deleted, cursor should be on the OTE instruction.
10. _____ Delete the OTE instruction.
11. _____ *Escape.*
12. _____ F6, *Delete Rung.*

EDIT THE *TON* INSTRUCTION

1. _____ Cursor to the TOF instruction.

Rung 2:0
The retentive timer retains its accumulated value when rung conditions become
false, a fault occurs, power is lost to the processor, or when you change the
processor from remote run to remote test mode and back to remote run mode.

```
|  I:1                                                        +RTO---------------+
|--] [-----------------------------------------------------+RETENTIVE TIMER ON+-(EN)-
|    0                                                      |Timer         T4:0+-(DN)
|                                                           |Time Base      1.0|
|                                                           |Preset          20|
|                                                           |Accum            0|
|                                                           +-----------------+
```

Rung 2:1
Timer timing bit is set when rung conditions are true and the accumulated
value is less than the preset value. This bit remains set until rung
conditions go false, or the done bit is set.

```
|  Timer
|  Timing
|  Input bit.
|
|    T4:0                                                                    O:2
|----] [--------------------------------------------------------------------( )--
|     TT                                                                      3
```

Rung 2:2
The timer enable bit is set when the timer instruction's rung is true.

```
|  Timer 4:0
|  Enabled
|
|    T4:0                                                                    O:2
|----] [--------------------------------------------------------------------( )--
|     EN                                                                      2
```

Rung 2:3
Timer four's done bit will energize output O:2/1 after the time delay of thirty
seconds.

```
|  T4:0                                                                      O:2
|--] [----------------------------------------------------------------------( )--
|   DN                                                                        1
```

Rung 2:4
Input I:1/1 is used to reset retentive timer T4:0.

```
|  I:1                                                                       T4:0
|--] [----------------------------------------------------------------------(RES)--
|   1
```

Figure 22-1   Edited *Timer_3* ladder diagram for lab exercise.

2. _____ Press F2, *Modify Rung*.
3. _____ Press F5, *Modify Instruction*.
4. _____ Notice cursor is blinking next to TON.
5. _____ Backspace to remove TON.
6. _____ Type RTO to edit timer to a retentive timer.
7. _____ Press enter. Notice timer address is displayed. You could change it if you Wanted to.
8. _____ Do not change address for this lab.
9. _____ Use the arrow keys to scroll through the different parameters.
10. _____ Change the preset from 30 to 20 by backspacing over it when it is highlighted in the instruction box and displayed next to the blinking cursor.
11. _____ Type *20*.
12. _____ Press *Enter*.
13. _____ Press *Enter*.

14. _____ Press *Escape*.
15. _____ Press *Accept Rung*, F10.
16. _____ Notice that the rung comment has disappeared.
17. _____ Enter the rung comment as illustrated in Figure 22-1 (previous page).
18. _____ When completed, press F8, *Accept / Exit*.
19. _____ Press F10, *Save Documentation*.
20. _____ Press *Escape*.
21. _____ Press F10, *Edit*.

ADD THE RESET INSTRUCTION

1. _____ Scroll to program, *Rung Four*.
2. _____ Press F4, *Insert Rung*.
3. _____ Press F4, *Insert Instruction*.
4. _____ Press F1, *Bit*.
5. _____ Press F1, *XIC*.
6. _____ Enter bit address *I:1/1*.
7. _____ Press *Enter*.
8. _____ Press F2, *Timer/ Counter*.
9. _____ Press F 8, *Reset*.
10. _____ Enter structure address as T4:0.
11. _____ Press *Enter*.
12. _____ Press *Escape*.
13. _____ Press F10, *Accept Rung*.
14. _____ Add documentation as illustrated in Figure 22-1.

Question

From the current screen, list the keys used to add the rung comment.

1. _____
2. _____
3. _____
4. _____
5. _____
6. _____
7. _____
8. _____

GO ON-LINE, DOWNLOAD, AND RUN THE PROGRAM

15. _____ Press F1, *Save / Go online*.
16. _____ Press F8, *YES*, when the software asks you if want to overwrite file.
17. _____ Press F8, *YES*, change processor to run mode.
18. _____ As program runs answer the following questions:

A. Energize I:1/0. Timer will run. Deenergize I:1/0. Does the RTO instruction retain the current accumulated value?_____

B. Energize I:1/0 again. Does the accumulated time value continue from where it left off?___

C. What would have happened to a TON instruction if you had cycled power to it as in question 1 above?_____

_____

D. Let the RTO run until the preset equals the accumulated value. With I:1/0 energized, what are the states of the status bits?

        DN = _____;

        TT = _____;

        EN = _____;

E. Deenergize I:1/0. Record status bit states:

        DN = _____;

        TT = _____;

        EN = _____.

F. Explain the difference in the answers for questions D and E._____

_____

_____

G. How do you reset the RTO?_____

_____

_____

23

PROGRAMMING A
COUNT-UP COUNTER

The typical counter counts from zero up to a predetermined value called the preset. As an example, to count from zero to 100, use a count-up counter. This exercise will step you through the procedure to develop a ladder program using a CTU instruction. Figure 23-1 on the next page illustrates the program we will develop.

THE LAB

1. _____ Copy the *Begin* file from your student floppy disk to your personal computer's hard drive.
2. _____ Rename the *Begin* file to *Count_up*.
3. _____ Press F8, *Monitor File*.
4. _____ Edit the count up program file.
5. _____ Press F4 to insert a new rung.
6. _____ Press F4 to insert instruction.
7. _____ Insert as XIC instruction with a bit address of I:1/1. The address you select should have a field wired switch so you can increment the counter when the program runs.
8. _____ To display the timer and counter instructions, press F2, *Timer/Counter*.
9. _____ Select the count-up counter by pressing F4, *CTU*. The count-up instruction should be on the right side of the ladder rung.
10. _____ "Enter Counter address" should be displayed in the lower left area of the screen.
11. _____ Type in C5:0 and press *Enter*. This address should be displayed in the counter instruction on the ladder rung.
12. _____ "Enter Present Value>", should be displayed in the lower left area of the screen.
13. _____ Type in *10* and press *Enter*.
14. _____ "Enter Accumulator value>" should be displayed.
15. _____ Type in *0* and press *Enter*. Entering the value zero in the accumulator causes our count to begin at zero.

Rung 2:0
Counter C5:0 is an up counter that will count up one decimal value each time
input I:1/1 transitions from false to true. The preset is 10 for this
application. When the preset value and accumulated value are equal, the
counter's done bit will become true, signaling the counter has reached the
preset value.

```
    I:1                                                  +CTU--------------+
 --] [-----------------------------------------------+COUNT UP           +- (CU) -
     1                                                |Counter       C5:0+- (DN)
                                                      |Preset          10|
                                                      |Accum            0|
                                                      +-----------------+
```

Rung 2:1
The done bit, C5:0/DN, also bit 13, will be true when the accumulated value is
equal to the preset.

```
    C5:0                                                                   O:2
 --] [----------------------------------------------------------------( )--
      DN                                                                 2
```

Rung 2:2
C5:0/CU is the counter count up enable status bit, bit 15. Whenever the
counter instruction rung is true, the count up enable status bit will also be
true.

```
    C5:0                                                                   O:2
 --] [----------------------------------------------------------------( )--
      CU                                                                 3
```

Rung 2:3
The C5:0/OV status bit will make this rung true when the counter's accumulated
value overflows the maximum count value.

```
    C5:0                                                                   O:2
 --] [----------------------------------------------------------------( )--
      OV                                                                 4
```

Figure 23-1   *Count_up* ladder program for exercise 23.

16. _____ This completes the counter instruction. Press *Escape* or *Enter* to return to the edit screen.
17. _____ Press F10 to accept the rung. This completes our first rung, rung zero.
18. _____ To start a new rung below the current rung, press F3, *Append Rung*.
19. _____ Insert an XIC instruction addressed as C5:/DN.
20. _____ Press *Enter*. The counter done bit will be used to control the output instruction when the accumulated value is equal to the preset.
21. _____ Add the output instruction, OTE, address as O:2/2.
22. _____ Accept the rung.
23. _____ Notice the software automatically appended the next rung.
24. _____ Program the count-up enable bit and enter the bit address C5:0/CU.
25. _____ Press *Enter*.
26. _____ Add an OTE instruction on this rung along with a valid output address. List the key strokes below.

      1. Press F1, *Bit*.

      2._____

      3._____

      4._____

      5._____

      6. *Accept Rung*, F10.

27. _____ Your rung should be accepted and a new rung added for you.
28. _____ Add an XIC instruction addressed as C5:0/OV, the counter overflow bit.
29. _____ List the key sequences to enter the overflow instruction:
    1. *Insert Instruction*, F4.

    2._____

    3._____

    4._____

    5._____
30. _____ Add a valid output instruction to this rung addressed as O:2/4.
31. _____ List the key sequences to accomplish this below:
    1. F1, *Bit*.

    2._____

    3._____

    4._____

    5._____
    6. Accept the rung.
32. _____ Press *Escape* to remove the automatically inserted rung.
33. _____ This completes this program.

SAVE AND GO ON-LINE

1. _____ Press F1, *Save / Go on line*.
2. _____ If you get the message: "File Already Exists. Overwrite File?"
3. _____ Press F8, *Yes*.
4. _____ If you get the message: "Processor program name does not match disk file program name. Continue with download?"
5. _____ Press F8, *Yes*.
6. _____ The message: "Program successfully restored to processor, change processor mode to run?" tells you that the program download from your personal computer to PLC memory was successful. Do you want to put the PLC into run mode?
7. _____ Press F8, *Yes* to put PLC into run mode.
8. _____ The run LED should be ON on your PLC processor.

TESTING YOUR PROGRAM

1. _____ Toggle the input switch that controls the counter input instruction. Each time the switch goes from OFF to ON, the counter accumulated value should increment by one.
2. _____ As you toggle the input to the counter, explain what happens to the status bits. When will these bits be true?

    Done bit_____

    Enable bit_____

    Overflow bit._____
3. _____ This counter's preset is programmed as 10. Explain what happens when the accumu-

lated value reaches the preset._____

_____

4. _____ How do status bits behave as the accumulated value increments pass the preset?
DN = _____; CU _____; OV _____.

MONITORING THE COUNTER DATA FILE

Let us take a look at the counter data table. We will monitor the counter file, file five. To monitor a data file:

1. _____ Move the cursor so it highlights the counter instruction. Only the instruction identification letters CTU are highlighted.

2. _____ Press F8, *Data Monitor,* to monitor the counter file five. Since our program has only one counter, only counter C5:0 will be displayed. Counter data should look similar to the following:

Address:	CV	CD	DN	OV	UN	UD	PRE	ACC
C5:0	0	0	1	0	0	0	10	12

The Counter Address is listed at the left. Status bits are listed in the center with preset and ACC on the right. Information we can get from monitoring this file includes status bit current status. This screen tells us the Done bit is set, and all other status bits are reset. The Preset is 10, and the Accumulated value is 12.

MONITORING OTHER DATA FILES

To monitor other data files, press the F7 key to move to the next data file; press the F8 key to move to the previous data file. The F5, *Specify Address* key allows you to specify a specific data file address that has been created for your program. As you look at the files, identify what is contained in each below:

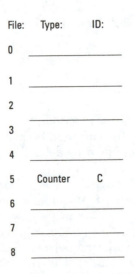

File:	Type:	ID:
0		_____
1		_____
2		_____
3		_____
4		_____
5	Counter	C
6		_____
7		_____
8		_____

Refer to your text for information to review data files and the type and format of data found in them. When finished return to monitor the counter data file screen.

OVERFLOWING THE COUNTER

In order to work with the overflow status bit, the counter must overflow. Rather than toggle your input toggle switch over 32,700 times, we will adjust the counter status file and adjust the accumulated valve.

1. _____ You should still be in the monitor counter file screen.
2. _____ Use the arrow keys to position the cursor to C5:0's accumulated value.
3. _____ Type in the value 32765
4. _____ Press *Enter*. The ACC Value should have changed to 32765.
5. _____ Press *Escape* to return to the ladder diagram. The counter on your ladder should have an accumulated value of 32765.
6. _____ Increment the counter using your input switch until the overflow status bit is set. Record the value when this happens.
7. _____ What happens to the accumulated value if you continue to increment?
8. _____ What happens the overflow status bit during this?
9. _____ To observe status bit operation as the counter goes from a negative value through zero into positive values, go to the data monitor screen, and change the counter accumulated value to a value like –2.
10. _____ Press *Escape* to return to the ladder program.
11. _____ Increment the counter through zero, into some positive value.
12. _____ Record what happens to the status bits as you do this. Done bit: _____; Overflow bit: _____.
13. _____ This counter will increment from –32768 back to zero and then on to 32767. To clear the counter to zero and reset the status bits, a reset instruction needs to be added to our ladder program.

ADD THE RESET INSTRUCTION

In order to reset the counter, a reset instruction with the same address as the counter must be programmed. We will now edit our program and add a reset.

1. _____ From the screen showing your ladder, press F10, *Edit*.
2. _____ The message, "Which Program Do You Want to Edit?"

> F1 Off line processor
> F3 Off line disk
> F7 Online edit

If you choose *Off line processor*, you will upload the current ladder and current data table values and the current processor file will be edited.

If you choose *Off line disk*, default data file data will be edited and downloaded when you are finished.

If your processor is a 5/03 or 5/04, the on-line edit option will be available. On-line editing is divided into two types. Program on-line editing allows editing the processor's program, with the processor in program mode, while on-line with your personal computer; run time on-line editing consists of editing the processor while it is in run mode, while on-line with your personal computer.

3. _____ Press F1, *Off line processor*. This selection will retain current data table values, including our counter's accumulated value.
4. _____ The message, "Saving Processor File to Disk. File already exists, overwrite file?" Select

F8, *Yes*. This will load the processor's ladder program on your personal computer's hard disk, overwriting the current version of this program.

5. _____ The message, "Processor in Run Mode, continue with upload?" Press F8, *Yes*.
6. _____ The current processor program will be uploaded to your hard drive. Your personal computer will now be off-line.
7. _____ Use the arrow keys to cursor to highlight the end rung.
8. _____ Press F4, *Insert Rung*. Program the rung as illustrated in Figure 23-2.
9. _____ Press F4, *Insert Instructions*.
10. _____ Enter the input instructions for this rung as illustrated in Figure 23-2.
11. _____ When finished entering the parallel branch, move cursor to the right power rail.
12. _____ Press F2, *Timer/Counter*.
13. _____ Press F8, *Reset*. This will enter a reset output instruction on your rung.
14. _____ Enter counter address C5:0.
15. _____ Press *Enter*.
16. _____ *Escape*.
17. _____ F10, *Accept Rung*.

```
Rung 2:0
Counter C5:0 is an up counter that will count up one decimal value each time
input I:1/1 transitions from false to true.  The preset is 10 for this
application.  When the preset value and accumulated value are equal, the
counter's done bit will become true, signaling the counter is done counting.
|   I:1                                                      +CTU--------------+        |
|--] [-----------------------------------------------------+COUNT UP          +- (CU) - |
|     1                                                     |Counter      C5:0+-(DN)    |
|                                                           |Preset         10|         |
|                                                           |Accum           7|         |
|                                                           +-----------------+         |

Rung 2:1
The done bit, C5:0/DN, also bit 13, will be true when the accumulated value is
equal to the preset.
|  C5:0                                                                       O:2  |
|--] [----------------------------------------------------------------------( )-- |
|   DN                                                                        2    |

Rung 2:2
C5:0/CU is the counter count up enable status bit, bit 15. Whenever the counter
instruction rung is true, the count up enable status bit will also be true.
|  C5:0                                                                       O:2  |
|--] [----------------------------------------------------------------------( )-- |
|   CU                                                                        3    |

Rung 2:3
The C5:0/OV status bit will make this rung true when the counter's accumulated
values overflows the maximum count value.
|  C5:0                                                                       O:2  |
|--] [----------------------------------------------------------------------( )-- |
|   OV                                                                        4    |

Rung 2:4
When either input I:1/2 or counter C5:0's done bit is true, the reset
instruction will reset the counters accumulated value to zero.
|     I:1                                                                    C5:0  |
|-+--] [--+---------------------------------------------------------------- (RES)-- |
| |   2   |                                                                         |
| | C5:0  |                                                                         |
| +--] [--+                                                                         |
|     DN                                                                            |
```

Figure 23-2 *Count_up* ladder program with reset instruction.

ADDING RUNG COMMENTS

1. _____ *Escape.*
2. _____ Enter rung comment for rung four as illustrated in Figure 23-2.
3. _____ When completed, press F8, *Accept/Exit.*
4. _____ Complete adding the documentation for all rungs, When finished entering the last documentation press F8, *Accept/Exit.*
5. _____ Press F10 to save documentation.
6. _____ Go on line and verify that your program functions properly.
7. _____ As your program runs fill in the table in Figure 23-3.

COUNTER STATUS BITS				
Status Bit	**Bit**	**When Set?**	**How Reset?**	**Sample Address**
Overflow (OV)	12			
Done (DN)				
Count-Up Enable (CU)				

Figure 23-3   Counter status bits.

8. _____ When you have verified that your program operates properly, save the program to your student floppy disk.
9. _____ If finished for this lab session, exit the program and shut down your computer.

24

PROGRAMMING A COUNT-DOWN COUNTER

The down-counter counts down over the range of +32,767 to –32,768. Each time the instruction sees a false-to-true transition, the accumulated value will be decremented by one count. This exercise will step you through editing your *Begin* ladder program and add a down-counter. Figure 24-1 on the next page illustrates the completed ladder program.

THE LAB

This exercise will step you through editing your *Begin* ladder program to add a down-counter. Figure 24-1 illustrates the completed ladder program.

1. _____ Load the *Begin* program from your student floppy disk to your personal computer's hard drive.
2. _____ Rename this program *Count_DN*.
3. _____ Enter rung zero from Figure 24-1. Make sure you use the *CTD* instruction.
4. _____ Accept the rung when finished entering the instructions.
5. _____ Enter rung one containing the done bit.
6. _____ Accept the rung when completed.
7. _____ Enter the reset rung.
8. _____ Accept the rung when completed.
9. _____ Add documentation to your program as in Figure 24-1.
10. _____ When your program is complete, save and go on-line.
11. _____ Put processor in run mode.
12. _____ As the program runs, decrement the counter through zero to a negative one. Answer the following questions on your observations.

QUESTIONS

1. When programming the down-counter, why do we set the accumulated value to twelve and the

```
Rung 2:0
The count down counter counts down from the accumulated value.  The
accumulated value is the beginning count that the counter will begin counting
down from. The accumulated value will decrement by one decimal value each time
the input instruction I:1/1 transitions from false to true.
 |  I:1                                                    +CTD--------------+      |
 |--] [-------------------------------------------------+COUNT DOWN      +-(CD)-    |
 |   1                                                  |Counter      C5:0+-(DN)   |
 |                                                      |Preset          0|        |
 |                                                      |Accum           9|        |
 |                                                      +-----------------+        |

Rung 2:1
The normally open done bit, C5:0/DN, also bit 13, will be true when ever the
accumulated value is greater than or equal to the preset.  Using a down
counter, the normally open done bit will be true any time the accumulated value
is greater than the preset.
 |  C5:0                                                                    O:2    |
 |--] [-----------------------------------------------------------------------( )--|
 |   DN                                                                       2    |

Rung 2:2
When ever input I:1/2 true, the reset instruction will reset the counters
accumulated value to zero.
 |  I:1                                                                     C5:0   |
 |--] [-------------------------------------------------------------------- (RES)--|
 |   2                                                                            |
```

Figure 24-1 Completed count-down counter program.

preset to zero, rather than the preset as twelve and the accumulated value as zero?_____

_____

_____

2. Explain what would happen if we programmed the accumulated value as zero, and the preset as twelve? This is the way the up-counter is programmed._____

_____

3. Each time the input instruction is true, the _____ will decrement one decimal value.

25

WORKING WITH A
COUNT-DOWN COUNTER

Assume you want to display on a visual display unit the number of remaining parts to be manufactured. Each lot is for 100 parts. The remaining parts to be build need to be displayed so workers can see how many parts are left to complete the lot. For this example the accumulated value will be set at 100, and the preset will be zero. Each time a part is completed and passes a sensor, the accumulated value will be decremented one decimal value.

THE LAB

We will edit Lab Exercise 24 and experiment with the application outlined above. The completed count-down ladder program, with a couple of modifications, could be interfaced to a visual display unit (commonly called an operator inferface device).

1. _____ Load the program from Exercise 24 and edit so the program matches Figure 25-1 (next page).

2. _____ Save and go on-line.

3. _____ Run program. Observe operation as counter decrements to zero. Record what you observe.

4. _____ Explain how the done bit operates._____

_____

5. _____ What is the function of the CD bit? Explain when this bit is set._____

_____

6. _____ How do you monitor the data file for the CTD C5:0 instruction? What key strokes do you use?_____

What information can you view now that you are in the data monitor screen?_____

_____

Rung 2:0
The count down counter counts down from the accumulated value. The accumulated value is the beginning count that the counter will begin counting down from. The accumulated value will decrement by one decimal value each time the input instruction I:1/1 transitions from false to true.

```
   I:1                                              +CTD-------------+
--] [-----------------------------------------------+COUNT DOWN      +- (CD) -
    1                                                |Counter   C5:0+- (DN)
                                                     |Preset       0|
                                                     |Accum      100|
                                                     +-------------+
```

Rung 2:1
The normally open done bit, C5:0/DN, also bit 13, will be true when ever the accumulated value is greater than or equal to the preset. Using a down counter, the normally open done bit will be true any time the accumulated value is greater than the preset.

```
   C5:0                                                            O:2
--] [------------------------------------------------------------( )--
    DN                                                             2
```

Rung 2:2
The normally closed done bit will be false whenever the accumulated value is greater than or equal to the preset. When using a down counter, the normally closed done bit will be false until the accumulated value transitions from zero to a negative one. As long as the accumulated value is less than zero this done bit will be true.

```
   C5:0                                                            O:2
--]/[------------------------------------------------------------( )--
    DN                                                             5
```

Rung 2:3
C5:0/CD is the counter count down enable status bit, bit 14. Whenever the counter instruction rung is true, the count down enable status bit will also be true.

```
   C5:0                                                            O:2
--] [------------------------------------------------------------( )--
    CD                                                             3
```

Rung 2:4
The C5:0/UN status bit will make this rung true when the counter's accumulated value underflows the minimum count value. Whenever the accumulated value reaches -32768 and wraps around to + 32767 the underflow bit will be true.

```
   C5:0                                                            O:2
--] [------------------------------------------------------------( )--
    UN                                                             4
```

Rung 2:5
When ever input I:1/2 true, the reset instruction will reset the counters accumulated value to zero.

```
   I:1                                                            C5:0
--] [---------------------------------------------------------- (RES)--
    2
```

Figure 25-1 Down-counter ladder program to be developed for this exercise.

Can you change the accumulated value from the data monitor screen? How do you do this?_____

7. _____ Change the accumulated value of C5:0 to –32,760.

8. _____ How do you return to monitor your ladder program?_____

9. _____ Observe as the counter decrements.

10. _____ Decrement the counter until the UN status bit is set.

			CTD STATUS BITS:			
Status Bit	**Bit Address**	**Counter at Rest, Rung False**	**I:1/1 is True. Counter Counting. Preset = 0 Accum = 10**	**I:1/1 is False. Counter Counting. Preset = 0 Accum = 0**	**Counter Counting Preset = 0 Accum = -x**	**Counter Wrap from -32,768**
Enable	C5:0/14					
Done						
Underflow						

Figure 25-2   Down-counter status bits.

11. _____ What is the accumulated value when the UN status bit is set?_____

12. _____ Decrement the counter one more time. What is the accumulated value?_____

13. _____ Is the UN status bit still set?_____

14. _____ What correlation to your CTD counter does the CU status bit have?_____

15. _____ If you wanted to send the decrementing count of your counter to a display for an operator to view, how would you handle the data within your program to send it from the counter to that display?_____

16. _____ If your counter's done bit is on continuously while the counter is decrementing to the ACCUM = PRE, how can you signal an output to turn on when the counter has indeed reached zero?_____

17. _____ Fill in the table in Figure 25-2 illustrating the bit conditions.

26

TIMER AND COUNTER PROGRAMMING

For this exercise, develop a timer and counter logic ladder program so as to get a better understanding of timer and counter programming and operation. Refer to the Advanced Programming Software Reference manual for additional information on timers and counters.

DEVELOP THE FOLLOWING PROGRAM

1. _____ Use input 03 to control TON timer element #6.
   A. Program the timer timing bit to control output 05.
   B. Program the timer done bit to control output 06.
   C. Program the timer enable bit to control output 07.
   D. Program timer to time up to thirty seconds.
2. _____ Also program the timer done bit as an input to counter element #2.
   A. Develop the rung of logic for an up-counter with a preset of ten.
   B. Develop a rung of logic so that when the counter hits its preset it will control output 01.
   C. Develop the logic required to reset the counter to zero after the preset has been reached. You will have to latch output 01 on so it will not go off.
3. _____ Develop a rung with a TOF timer controlling output 02.
4. _____ Develop a rung of logic with a CTD counter controlling output 08. Program this counter so it will count down from ten to zero. When the accumulated value is zero output eight will energize. Also include a way to reset the counter.

QUESTIONS

1. When starting development of a new program, what steps do you have to go through in the software before you can actually add the first rung of logic?_____

_____

2. From the main menu screen, what keys do you have to press to get to the edit screen where

you will add your new rungs and instructions?_____

3. Based on watching the operation of the TON timer, explain briefly how this operates._____

_____

4. Based on watching the operation of the TOF timer, explain briefly how this operates._____

_____

_____

5. What is the basic difference between the operations of the TON and the TOF timers?_____

_____

6. Explain how the timer timing bit operates._____

_____

7. Explain how the done bit operates._____

_____

8. What does the timer enable bit tell you?_____

_____

9. List an example of each of the possible addressing formats for the done bit for timer element

2._____

_____

10. List an example of each of the possible addressing formats for the timer timing bit for TON

element 3._____

_____

11. List an example of each of the possible addressing formats for the accumulated value for T4.

_____

_____

12. List an example of each of the possible addressing formats for the preset value for timer zero.

_____

_____

13. Make a grid showing the operation of the TT, EN, and DN bits (a) at the time the TON timer
    is at zero and not energized, (b) with the timer energized and incrementing toward the preset,
    and (c) at the preset.

14. What is the counting range of a counter?_____

15. What information does bit 11 and 12 in word zero of your counter element provide?_____

_____

16. If output O:2.0/6 is to energize when a counter reaches its preset value, what is going to control

    the OTE instruction to cause it to turn on or off in relation to the counter's state?_____

_____

17. Explain how the CTU operation differs from the CTD._____

18. What does the preset represent while using a CTD counter?_____

19. How is your CTD counter reset?_____

20. Explain how a counter counts in relation to its input pulses._____

21. Explain how the count-down underflow bit works._____

22. Explain how the done bit works for a CTD counter._____

23. Explain how the count-down enable bit works._____

24. List an example of each of the possible addressing formats for the count-up enable bit for counter element 3._____

25. List an example of each of the possible addressing formats for the done bit for counter element 3._____

26. List an example of each of the possible addressing formats for the accumulated value for counter element 3._____

27. Save this program on your floppy disk with the file name: *Count_A.*

27

FORCING INPUTS AND OUTPUTS

The PLC has a troubleshooting feature called "forcing" that enables the troubleshooter to override and force the actual status of input data file bits to ON or OFF, or override the processor logic and status of the output status file bits by forcing outputs ON or OFF.

Forcing an input ON or OFF can be used to test the processor logic and output interaction. When used as a troubleshooting tool, forcing an input signal can be used to verify program operation by providing a known input signal or signals.

Forcing an output ON or OFF is a troubleshooting technique allowing you to enter a ON or OFF output status bit and verify operation of the PLC output circuits, output module I/O point or points, external circuitry, and the output field hardware device.

Even though forcing inputs or outputs can be an invaluable troubleshooting tool, extreme care must be taken to insure safety. Keep in mind that there is a possibility of personal injury and machine damage caused by unexpected machine operation as the result of forcing an input or output ON or OFF. Make sure you understand the effects on machine operation before forcing input data file bits, or external output circuits.

FORCE FILE PROTECTION

The SLC 500 5/03 and 5/04 processors have a force file protection option that can be applied to a file before saving the file, or testing edits. The force file protection feature will protect a file against forces by individuals in the field where safety or machine damage is a concern.

THE LAB

In this section we will work with forcing. We want to force input I:1/2 ON in rung 2 of the force program we are about to create.

1. _____ Start with your *Begin* program.
2. _____ Rename *Begin* program *Force*.

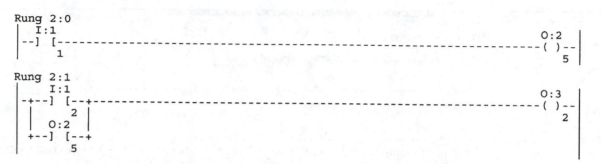

Figure 27-1 Force program.

3. _____ We will investigate an alternative method of entering the program illustrated in Figure 27-1.
4. _____ Enter the program logic using the following key strokes:
5. _____ From the edit screen, F4, *Insert Rung.*
6. _____ To insert the XIC instruction, type *XIC* at the blinking cursor.
7. _____ Press *Enter.*
8. _____ Enter bit address: *I:1/1.*
9. _____ Press *Enter.*
10. _____ Type in *OTE* at the blinking cursor.
11. _____ Press *Enter.*
12. _____ Enter bit address > *O:2/5.*
13. _____ Accept the rung.
14. _____ Enter rung one in the same manner.
15. _____ Save and go on-line.
16. _____ Run program
17. _____ Make sure the input field device for I:01/2 is OFF.
18. _____ Put the cursor on the I:1/2 instruction.
19. _____ Press *Force,* F9.
20. _____ Press *ON,* F2.
21. _____ Press F7, to enable force.
22. _____ Press F8, *Yes.*
23. _____ Do you see any change in the status lights on the processor?
24. _____ Do you see any change in your computer's display near the I:1/2 instruction?
25. _____ Is output 0:3/2 ON? _____
26. _____ Press F8 to disable the force.
27. _____ Press F8, *Yes,* to disable forces.
28. _____ Does output 0:3/2 go off? _____
29. _____ Was there any change to I:1/2 ? _____
30. _____ Press F7 to re enable the force.
31. _____ Press F8, *Yes* to verify you want to enable force.
32. _____ Press F4 to remove all forces.
33. _____ F8, to acknowledge.
34. _____ Cursor to the 0:2/5 parallel input instruction on rung 1.
35. _____ Force this input ON. From the edit file screen: what are the four key strokes that take

you through the procedure?_____
36. _____ At this point, only 0:2/5 should be forced ON.
37. _____ Cursor to I:1/2. Force this ON.
38. _____ If we have a number of I/O forced ON in different places in our program and want to

look to see which inputs and outputs were ON—press F9, *Monitor Inputs.* The input status file is displayed.

39. _____ Explain what you see in the input status file._____

_____

40. _____ What does this signify?_____

_____

41. _____ Look at the input module in slot one of your SLC chassis. Is the LED indicator light on for I:1/2 ? Why, or why not?_____

_____

42. _____ To look at the output status file there are two options:

   A. Press F5, *Specify Address.*
   B. Enter the Force table address. Type in the letter *O,* for output status file.
   C. Press *Enter.*
   D. The output status file is displayed with *ones* displayed for outputs forced ON.

43. _____ A second way to display the output status file is to press *Escape.*
44. _____ Press F9, *Force.* This brings you back to the Force screen.
45. _____ Press F10, *Outputs.* Forced outputs will be displayed.
46. _____ You can press F6, *Data Monitor* to display the entire output status file.
47. _____ You can press F7 to view the next file, which is input file one.
48. _____ Press F8 to move back one file to output file, file zero.
        NOTE: By pressing the F7, *Next File,* or the F8, *Previous File,* all data files can be stepped though and displayed.
49. _____ Press *Escape.*

50. _____ To disable the forces, what keys do you press?_____
51. _____ To remove the force from instruction I:1/2, cursor to the instruction and press F3, to remove the selected force, or since there are multiple forces, press F4, *Remove All.*
52. _____ Press F8, to verify you wish to do this.
53. _____ Press *Escape.*
54. _____ This will return you to the monitor program screen.
55. _____ If you are finished for today, save program to your floppy disk.

SUMMARY

We experimented with an easier method to add ladder instructions by directly keying the instruction abbreviation. This method of instruction entry can be used with any instruction, assuming you can remember the instruction abbreviation. Practice this alternate method of instruction entry as you proceed through the upcoming exercises.

We also worked with forcing I/O. Forcing can be a valuable troubleshooting tool if used correctly. Remember, *forcing should be used with extreme caution in a real production environment.* Make sure your particular company allows forcing of PLC logic before proceeding. And remember, *safety first!*

28

ADVANCED
EDITING

The PLC has cut, copy, and paste functions which are classified as advanced editing. These editing functions can be used to cut entire rungs, or groups of rungs. A rung can be copied to another portion of the current program, or even pasted into another program. The paste function is used to paste a rung or group of rungs into position after they have been cut or copied.

In this section we will work with the cut, copy and paste features of the software package. We will start with our *Force* program from the last exercise. We will use the cut and paste, along with the copy-and-paste, functions to move rungs around within our program. After practicing moving rungs within the *Force* program, we will copy rungs from the *Force* program to another program.

THE LAB

1. _____ Start with the *Force* program from the previous exercise.
2. _____ From the edit screen we will cut rungs from the program.
3. _____ Cursor to rung zero.
4. _____ Press F8, *Advanced Editing*.
5. _____ Press F1 to select the rung.
6. _____ Press F2 to cut the rung from the program.
7. _____ To paste the cut rung to another location in the program, move the cursor to the *End* ladder rung.
8. _____ Press F4, *Paste*.

COPY RUNGS TO OTHER LOCATIONS WITHIN THE SAME PROGRAM

9. _____ To copy a rung from one part of the program and paste a copy of the original rung to another location in the program, press F1, *Select*.
10. _____ Use the arrow keys to select both rungs.
11. _____ Press F3, *Copy*.
12. _____ Move cursor to the *End* rung.
13. _____ Press F4, *Paste*.

CUT OR COPY RUNGS FROM ONE PROGRAM TO ANOTHER

14. _____ Select rungs zero and one from the *Force* program.
15. _____ F3, *Copy*.
16. _____ *Escape.*
17. _____ *Escape.*
18. _____ F3, *Exit.*
19. _____ *Escape.*
20. _____ F7, *File Options.*
21. _____ Load *Begin* program.
22. _____ Edit *Begin* program.
23. _____ Go to *Advanced Editing.*
24. _____ F4, *Paste.* The two rungs copied from the *Force* program should be copied to the new *Begin* program.

QUESTIONS

1. When doing Advanced Editing, the select option:
   A. Allows you to select a single instruction you wish to edit.
   B. Allows you to select a single rung you wish to edit.
   C. Allows you to select multiple rungs you wish to edit.
   D. A and B are correct
   E. B and C are correct
   F. A, B and C are correct.

2. When editing, the Cut option:
   A. Cuts, or deletes, a selected instruction from your ladder.
   B. Cuts, or deletes, a selected rung or rungs from your ladder.
   C. Cuts, or deletes, selected files from your ladder.
   D. Cuts, or deletes, selected status bits from your ladder.
   E. Cuts, or deletes, selected reset instructions from your ladder.

3. The Copy function allows you to:
   A. Copy an instruction from a selected ladder rung and put a copy of it on another rung.
   B. Copy a single rung from your current program to another location in the same program.
   C. Copy selected rung(s) from your program to another location in that same program.
   D. Copy selected rung(s) from your current program to another location in the same program or another program.

4. The Paste function allows you to:
   A. Paste an instruction to be copied into the selected position.
   B. Paste the rung cut from a program into a new position in the current program or a selected program.
   C. Paste selected files into your ladder.
   D. Paste selected status bits to your ladder.
   E. A, B, and C are correct.
   F. A and B are correct.

5. After cutting from a program a rung that you want to paste to another location in the same program, the cursor is positioned:
   A. On the end rung of your ladder. You enter the rung position where you want to place the cut rung and press F4, *Paste*.
   B. On the rung above the position where you want to paste the rung, Press F4, *Paste*.
   C. On the rung below the position where you want to paste the rung, Press F4, *Paste*.
   D. On the rung where you want to paste the rung, Press F4, *Paste*. The pasted rung will take the place of the cursored rung, which will move down one rung position, as will all succeeding rungs.

29

TIMER AND COUNTER PROGRAMMING PRACTICE

This lab exercise will give you practice developing a program using timers and counters for a specific application. When finished developing the program, you will download it and verify its operation.

Your task is to develop a program that will control an auxiliary cooling fan on a motor controlled by a variable frequency drive. Running time of the motor must also be tracked for maintenance.

INTRODUCTION

One of the maintenance people has just added an induction motor and a variable-speed drive to one of the manufacturing lines. The motor will run at a very low speed during some portions of the manufacturing cycle. To avoid possible motor overheating problems, an auxiliary fan has been installed to keep the motor cool. Develop a program to control the motor auxiliary cooling fan and maintenance timers as specified below.

Rather than adding the new rungs directly to the production program, develop this new portion of the program on your desktop computer, and verify operation. When the program has been tested for proper operation, it will be pasted into the current production program, much as you did in the last exercise.

PERTINENT INFORMATION

1. Three-wire control through a momentary start-stop push button station will control the operation of a motor. Configure the start/stop station as inputs to our PLC.
2. The motor is controlled by a variable frequency drive (VFD) because the motor speed needs to be varied depending upon what process is being performed. The PLC will send a signal to start the VFD.
3. For this exercise we will not deal with programming the drive for varying the motor speed, or start/stop.
4. Since some manufacturing processes require the motor to run at very low speeds, to eliminate possible motor overheating problems an auxiliary cooling fan has been installed for motor cooling.

5. Program the PLC control for the cooling fan so that when the motor is turned off, the fan continues to run for thirty minutes.
6. The maintenance department needs to perform maintenance after every 500 hours of motor running time. When 500 hours of running time have elapsed, a pilot light will illuminate on the control panel alerting the operator that it is maintenance time. If maintenance is not completed by 550 hours, an alarm bell will sound.

PROGRAM DEVELOPMENT SUGGESTIONS

1. For this exercise, develop the start /stop logic along with the timing sequences and outputs.
2. A timer will be needed that will not forget its accumulated elapsed time when the motor is

   turned off, or as a result of a power failure. What sort of timer performs like this?_____

   _____
3. The easiest way to track time in hours is to build a clock consisting of a timer that will run for one hour; after one hour the timer will send a pulse to a counter which will track the elapsed hours. To determine the timer's preset, use a one-second time base. One second times 60

   seconds equals one minute. Sixty minutes equal one hour. What is your preset?_____
4. Our counter will count the running hours of the motor. We have two points in time that we wish to keep track of. First, when the motor has run 500 hours we need the pilot light to go on. Second, if at the 500-hour mark maintenance was not performed and the maintenance timer reset, an alarm bell is to sound when the motor has run 550 hours. For this exercise, use one counter for 500 hours (C5:0), and another (C5:1) for the 550-hour count.
   > HINT: To enable test viewing the completed running program, you may wish to lower your presets from thirty minutes for the after-shutdown of the motor fan running time, to thirty seconds. Also consider shortening the presets for the motor run time portion of your program to, say, fifty seconds and fifty-five seconds respectively.
5. Don't forget to include the proper reset instructions.
6. Develop program, run and test for proper operation.

QUESTIONS

1. What timer instruction did you use for control of the motor fan?_____

2. Explain why you selected this timer instruction._____

   _____

3. How did you trigger the pilot light off the counter accumulated value of 500 hours?_____

   _____

SUMMARY

In this exercise we learned how to build a clock to lengthen our timing cycle past the one-second time base provided by a single timer. As we developed our program we learned how to use the proper timer instruction to track time as it accumulates through many machine cycles, through power failures, or if the main disconnect is opened.

We learned how we can keep an output running for a specified time even after the output to a specific device, in this case a slow running motor, has been de-energized.

In a future lab exercise we will modify this exercise and use one counter with comparison instructions.

30

COMPARISON
INSTRUCTIONS

This exercise will introduce you to incorporating comparison instructions to trigger events at counter accumulated values other than when the accumulated value and preset are equal.

INTRODUCTION

Comparison instructions are input instructions that test two values to determine if the instruction will be true or false. Comparison instructions include equal, not equal, less than, less than or equal, greater than, greater than or equal, masked comparison for equality, and the limit test, all to test one value against another. Most comparison instructions have two parameters, source A and source B. Typically source A must be an address, while source B can be either a data file address or a constant. As an example, using an *Equal* instruction, source A could be a counter's accumulated value, and source B could be a constant, such as 10. When the counter accumulated value is equal to 10, the instruction will be true. On the other hand, if source B is an address, say N7:0, then the value stored in N7:0 is used in the comparison process.

THE LAB

For this exercise we will program a counter on rung zero, and then program many of the comparison instructions, each controlling an output instruction. As we run the program, we will increment the counter and observe how the different comparison instructions behave. The ladder program for this exercise is illustrated in Figure 30-1 on pages 142–143.

1. _____ Copy the *Begin* program to your hard disk.
2. _____ Rename the *Begin* program to *Comp_1*.
3. _____ Go to *Offline Programming / Documentation*.
4. _____ Use the *Change File* function and select the *Comp_1 file*.
5. _____ Monitor and edit the file.
6. _____ Program rung zero as illustrated in Figure 30-1.

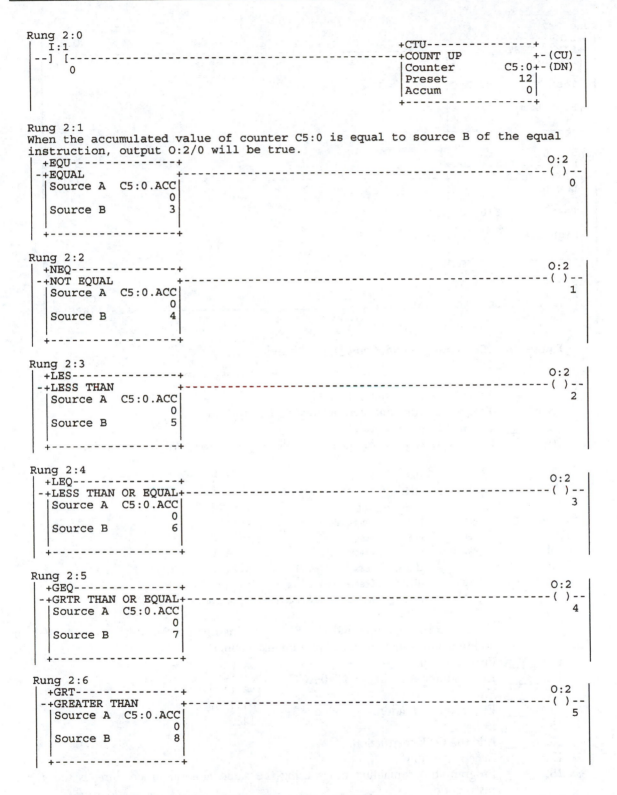

```
Rung 2:0
   I:1                                                          +CTU---------------+
--] [-------------------------------------------------------+COUNT UP           +-(CU)-
   0                                                         |Counter       C5:0+-(DN)
                                                             |Preset          12|
                                                             |Accum            0|
                                                             +------------------+

Rung 2:1
When the accumulated value of counter C5:0 is equal to source B of the equal
instruction, output O:2/0 will be true.
                                                                              O:2
   +EQU--------------+                                                         | |
-+EQUAL           +-----------------------------------------------------------( )--
   |Source A  C5:0.ACC|                                                         0
   |               0|
   |Source B        3|
   |                 |
   +-----------------+

Rung 2:2
   +NEQ--------------+                                                         O:2
-+NOT EQUAL       +-----------------------------------------------------------( )--
   |Source A  C5:0.ACC|                                                         1
   |               0|
   |Source B        4|
   |                 |
   +-----------------+

Rung 2:3
   +LES--------------+                                                         O:2
-+LESS THAN       +-----------------------------------------------------------( )--
   |Source A  C5:0.ACC|                                                         2
   |               0|
   |Source B        5|
   |                 |
   +-----------------+

Rung 2:4
   +LEQ--------------+                                                         O:2
-+LESS THAN OR EQUAL+----------------------------------------------------------( )--
   |Source A  C5:0.ACC|                                                         3
   |               0|
   |Source B        6|
   |                 |
   +-----------------+

Rung 2:5
   +GEQ--------------+                                                         O:2
-+GRTR THAN OR EQUAL+----------------------------------------------------------( )--
   |Source A  C5:0.ACC|                                                         4
   |               0|
   |Source B        7|
   |                 |
   +-----------------+

Rung 2:6
   +GRT--------------+                                                         O:2
-+GREATER THAN    +-----------------------------------------------------------( )--
   |Source A  C5:0.ACC|                                                         5
   |               0|
   |Source B        8|
   |                 |
   +-----------------+
```

Figure 30-1   Compare program *Comp_1. (continues)*

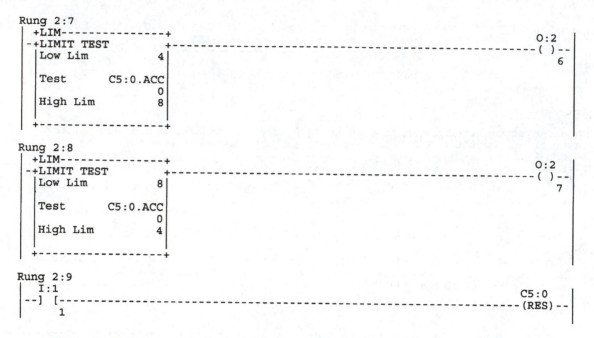

```
Rung 2:7
  +LIM---------------+                                                       O:2
 -+LIMIT TEST        +--------------------------------------------------------( )--
  |Low Lim          4|                                                        6
  |
  |Test      C5:0.ACC|
  |                 0|
  |High Lim         8|
  +------------------+

Rung 2:8
  +LIM---------------+                                                       O:2
 -+LIMIT TEST        +--------------------------------------------------------( )--
  |Low Lim          8|                                                        7
  |
  |Test      C5:0.ACC|
  |                 0|
  |High Lim         4|
  +------------------+

Rung 2:9
   I:1                                                                       C5:0
 --] [---------------------------------------------------------------------(RES)--
    1
```

Figure 30-1   Compare program *Comp_1. (continued)*

7. _____ Accept the rung when completed.
8. _____ To program rung one start by pressing F3, *Append Rung*.
9. _____ F4, *Insert Instruction*.
10. _____ F4, *Comparison Instructions*. The following selections will be displayed:

F2	LIM	Limit Test
F3	MEQ	Masked Comparison for Equal
F4	EQU	Equal
F5	NEQ	Not Equal
F6	LES	Less Than
F7	GRT	Greater Than
F8	LEQ	Less Than or Equal
F9	GEQ	Greater Than or Equal
F10	Others	Returns you to instruction selection screen

Refer to your text and the SLC 500 Instruction Set Reference Manual for additional information on comparison instructions.

11. _____ Press F4, *Equal*.
12. _____ Enter source A address > C5:0.ACC.
13. _____ *Enter.*
14. _____ Enter source *B* address or value > 3.
15. _____ *Enter.*
16. _____ Add the OTE instruction.
17. _____ Accept the rung.
18. _____ Program the remaining rungs using the same general proceedure as the *Equal* instruction.
19. _____ When finished programming all rungs, save and go on line.
20. _____ As your program runs, observe the behavior of each comparison instruction.

21. _____ As you increment the counter and observe each comparison instruction, go back and add explanatory rung comments for each comparison instruction rung.

22. _____ Complete the question section below.

23. _____ When completed, save program to your floppy disk.

QUESTIONS

1. When is the *equal* instruction true?_____

2. When is the *not equal* instruction true?_____

3. As the counter increments, when is the *less than* instruction true?_____

_____

4. When is the *less than or equal* instruction true?_____

5. When is the *greater than or equal* instruction true?_____

6. When is the *greater than* instruction true?_____

7. Explain the operation of the *limit test* instruction on rung seven._____

_____

8. Explain the operation of the *limit test* instruction on rung eight._____

_____

31

COMPARISON INSTRUCTIONS AND MULTIPLE COUNTER PRESETS

This exercise will introduce you to incorporating comparison instructions to trigger events off of counter accumulator values other than when the accumulated value and preset are equal.

In Exercise 29 we used two counters, one to trigger our maintenance reminder pilot light after 500 hours, and a second counter to energize the maintenance alarm at 550 hours. We can use a data handling instruction such as the *equal* instruction to test the counter accumulator for a value equal to a predetermined value, other than the counter's preset value, and then initiate some action when the *equal* instruction is true. In this exercise we will use the *equal* instruction to test for our counter's accumulated value to be 500. When the test value of 500 and the counter accumulated value are equal, at 500 hours, the equal instruction will be true. With the *equal* instruction being true, we will latch the maintenance pilot light on. Figure 31-1, on pages 147–148, illustrates a sample ladder with the *equal* instruction added to an example of Lab Exercise 29's ladder program.

THE LAB

We will be editing Lab Exercise 29's ladder diagram. With the program loaded in your personal computer edit the program as outlined below:

PART I

1. _____ Delete counter C5:1 from your ladder program.
2. _____ Press F4, *Insert Rung.*
3. _____ Press F4, *Insert Instruction.*
4. _____ Press F4 To choose the group of compare instructions.
5. _____ Press F4, *Equal,* to place the *equal* instruction on your ladder rung. Fill in the following parameter information.

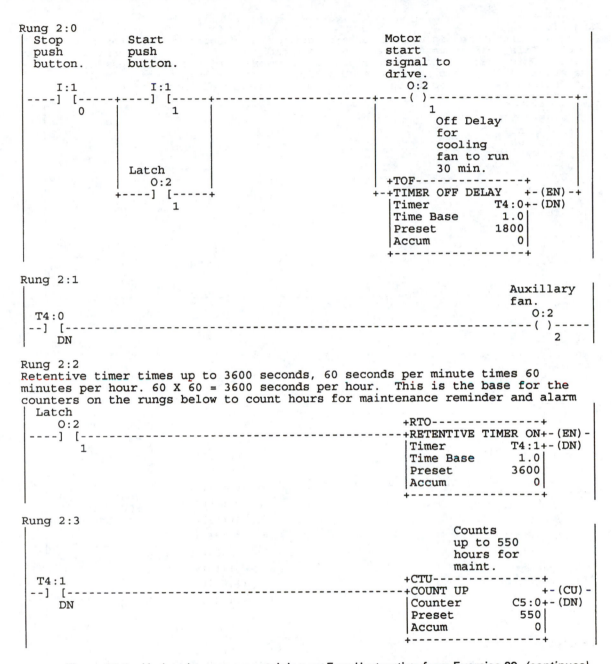

```
Rung 2:0
| Stop             Start                              Motor
| push             push                               start
| button.          button.                            signal to
|                                                     drive.
|     I:1              I:1                              O:2
----] [-----+----] [-----+------------------------+----( )-----------------+-
      0     |          1 |                         |    1
            |            |                              Off Delay
            |            |                              for
            |            |                              cooling
            |            |                              fan to run
            |   Latch    |                              30 min.
            |    O:2     |                         +TOF---------------+
            +----] [-----+                       +-+TIMER OFF DELAY    +-(EN)-+
                 1                               | |Timer          T4:0+-(DN)
                                                 | |Time Base       1.0|
                                                 | |Preset         1800|
                                                 | |Accum             0|
                                                 | +------------------+

Rung 2:1
|                                                           Auxillary
|                                                           fan.
|  T4:0                                                      O:2
--] [---------------------------------------------------------( )-----|
     DN                                                      2

Rung 2:2
Retentive timer times up to 3600 seconds, 60 seconds per minute times 60
minutes per hour. 60 X 60 = 3600 seconds per hour.  This is the base for the
counters on the rungs below to count hours for maintenance reminder and alarm
|  Latch                                               +RTO---------------+
|   O:2                                              +-+RETENTIVE TIMER ON+-(EN)-
----] [---------------------------------------------+ |Timer          T4:1+-(DN)
     1                                                | |Time Base       1.0|
                                                      | |Preset         3600|
                                                      | |Accum             0|
                                                      | +------------------+

Rung 2:3
|                                                      Counts
|                                                      up to 550
|                                                      hours for
|                                                      maint.
|  T4:1                                               +CTU---------------+
--] [------------------------------------------------+COUNT UP           +-(CU)-
     DN                                               |Counter        C5:0+-(DN)
                                                      |Preset          550|
                                                      |Accum             0|
                                                      +------------------+
```

Figure 31-1 Updated program containing an *Equal* instruction from Exercise 29. *(continues)*

Rung 2:4
The equal instruction is used to obtain action from counter C5:0 at the count
of 500. This enables us to effectively have multiple presets from a single
counter. This counter has a preset of 550, however using the equal instruction
also provides us with a second preset at 500 counts. Multiple comparison
instructions can be used in this manner to obtain multiple pick off points as
the counter accumulates.

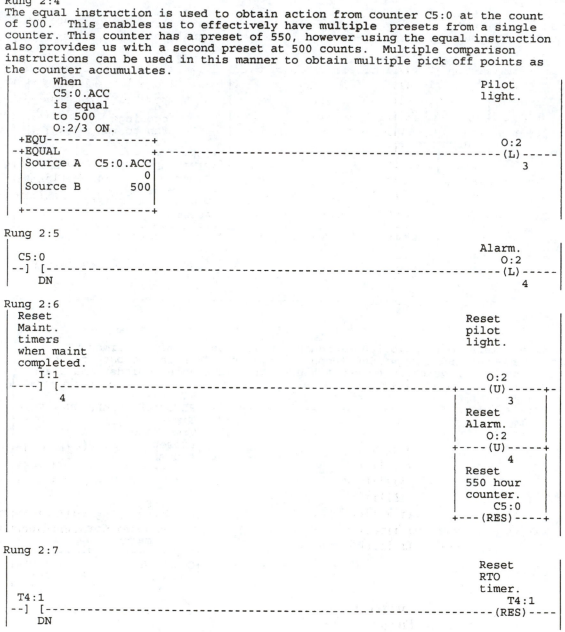

Figure 31-1   Updated program containing an *Equal* instruction from Exercise 29. *(continued)*

        Enter Source *A* address > *C5:0.ACC.*

        Enter Source *B* address or value > *500.*

6. _____ To enter the rung's output, start by Pressing F1, *Bit.*

7. _____ Press F4, *OTL* (Output Latch).

        Enter bit address> *O:2/3.*

8. _____ Make any other needed modifications to your program.

9. _____ Go on line, download and run the program.

PART II

The maintenance people state they have trouble noticing when the motor maintenance pilot light turns on. As a result, maintenance is not being completed on a timely basis. You task is to find a way to solve this problem.

After talking with the maintenance people, you decide to modify the PLC program so the maintenance reminder light, instead of simply coming on, will flash on and off when motor running time exceeds 500 hours. You will incorporate a *greater than or equal to (GEQ) instruction* to test when Counter C5:0.ACC is greater than or equal to 500 hours. The GEQ instruction will be true when the counter accumulated value is 500 hours or greater. Develop program logic so that when the GEQ instruction is true, the pilot light will flash on and off.

A timer could be used to accomplish this; however, there is an easier way. The SLC 500 has a free-running clock in the status file at address S:4. The free-running clock is simply 16-bit word that is incremented by the CPU at a rate of 20 milliseconds for the LSB. Figure 31-2 illustrates the time in milliseconds for each bit of S:4's word.

Bit Number	Milliseconds	Bit Number	Milliseconds
0	20	8	5120
1	40	9	10,240
2	80	10	20,480
3	160	11	40,960
4	320	12	81,920
5	640	13	163,840
6	1280	14	327,680
7	2560	15	655,360

Figure 31-2   Status file S:4's time value corresponding to each bit position of the sixteen-bit word.

You can address any bit from S:4's sixteen bits in your user program and have the addressed ladder instruction toggle at the associated predetermined ON/OFF time. For this exercise we will arbitrarily select addressing S:4/6 with an XIC instruction that would cause the XIC instruction to toggle on and off every 1280 milliseconds.

Modify your program so it is similar to the program in Figure 31-3 on the next page for control of the maintenance pilot light. Refer to your SLC 500 Instruction Set Reference Manual for more information on the status file, along with S:4.

Developing the GEQ Rung

1. _____ Delete the rung containing the *EQU* instruction.
2. _____ F4, *Insert Rung*.
3. _____ F4, *Insert Instruction*.
4. _____ F4 to select the compare group of instructions.
5. _____ F9, *GEQ*.
   Enter Source A address> *C5:0.ACC*.
   Press *Enter*
   Enter Source B address or value> *500*
   Press *Enter*
6. _____ F1, *Bit*.
7. _____ F1, *XIC*.
   Enter bit address> *S:4/6*
8. _____ F1, *Bit*.

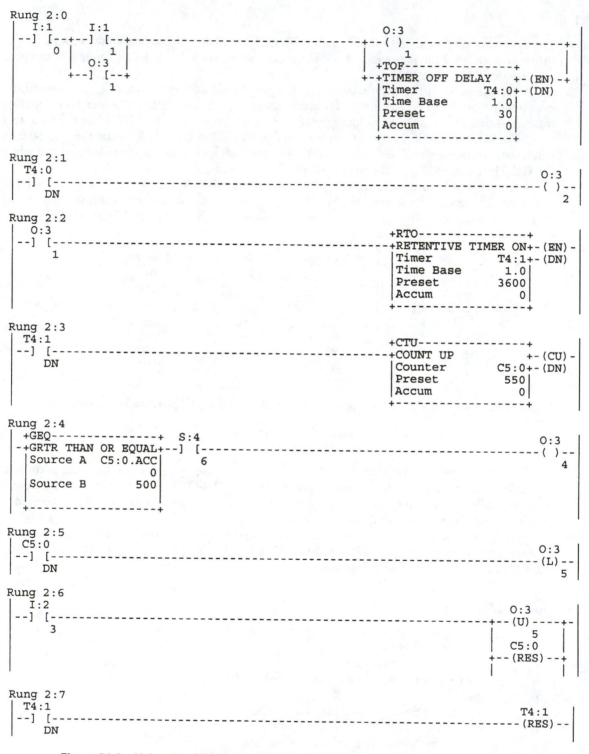

```
Rung 2:0
   I:1      I:1                                        O:3
--] [--+--] [--+----------------------------------+--( )------------------------+-
     0  |    1   |                                 |   1
        | O:3    |                              +--+TOF---------------+
        +--] [--+                               +--+TIMER OFF DELAY   +--(EN)-+
             1                                     |Timer      T4:0+--(DN)
                                                   |Time Base    1.0|
                                                   |Preset        30|
                                                   |Accum          0|
                                                   +----------------+

Rung 2:1
   T4:0                                                               O:3
--] [------------------------------------------------------------------( )--
     DN                                                                 2

Rung 2:2
   O:3                                                 +RTO---------------+
--] [-----------------------------------------------+-+RETENTIVE TIMER ON+-(EN)-
     1                                                |Timer      T4:1+--(DN)
                                                      |Time Base    1.0|
                                                      |Preset      3600|
                                                      |Accum          0|
                                                      +----------------+

Rung 2:3
   T4:1                                                +CTU---------------+
--] [-----------------------------------------------+-+COUNT UP        +--(CU)-
     DN                                                |Counter    C5:0+--(DN)
                                                       |Preset       550|
                                                       |Accum          0|
                                                       +----------------+

Rung 2:4
  +GEQ---------------+  S:4                                           O:3
-+GRTR THAN OR EQUAL+--] [------------------------------------------( )--
   |Source A  C5:0.ACC|   6                                            4
   |               0|
   |Source B     500|
   |                |
   +----------------+

Rung 2:5
   C5:0                                                               O:3
--] [------------------------------------------------------------------(L)--
     DN                                                                 5

Rung 2:6
   I:2                                                               O:3
--] [----------------------------------------------------------+--(U)----+-
     3                                                         |   5     |
                                                               | C5:0    |
                                                               +--(RES)--+
                                                               |         |

Rung 2:7
   T4:1                                                             T4:1
--] [----------------------------------------------------------------(RES)--
     DN
```

Figure 31-3   Using the GEQ in conjnction with S:4 to control the maintenance reminder pilot light.

9. _____ F3, *OTE.*
    Enter bit address> O:3/4
10. _____ Press *Escape.*
11 _____ F10, *Accept Rung.*
12. _____ Go on line, download and run and test your program.
13. _____ Save program to your floppy.

QUESTIONS

1. How does the *GEQ* instruction's control of an output differ from the *EQU* instruction?_____
_____

2. What does it mean when we say that S:4's LSB is equal to 20 milliseconds?_____
_____

3. What instruction would you use to create the following application?

   A. Output O:3/2 is true when counter C5:24's accumulated value is between zero and fifty.___
   _____

   B. Output O:3/5 is true when counter C5:24's accumulated value is equal to 100._____
   _____

   C. Output O:3/6 is true when counter C5:24's accumulated value is above 100._____
   _____

   D. Output O:3/7 is true when counter C5:24's accumulated value is not 125._____
   _____

SUMMARY

This Lab Exercise provided practice using two comparison instructions, the *Equal* and the *Greater Than or Equal.* Even though other data handling instructions were not incorporated into this lab exercise, they all program similar to the two used here.

Practicing incorporating a status file word's bits into the ladder program, in order to use the free-running clock function, was accomplished by using status bit S:4. Using this bit was an easy way to program our pilot light to flash on and off to get the attention of the maintenance people and remind them it was time to perform needed maintenance on the motor.

32

THE
COPY INSTRUCTION

The copy instruction copies blocks of data from one data file location to another data file location. This is called a file-to-file transfer. Data in the original location, called the source, is left intact while a copy if the source information is copied to the designated destination. Copying a group of words of like data is called copying a *user-defined file*. Data can be copied to another location either within the same data file, or to another data file. A user-defined file could contain recipe data of all the ingredients needed to make a batch of chocolate chip cookies. The copy instruction could be used to copy the chocolate chip recipe into the proper PLC data table so that chocolate chip cookies can be made when the current batch of peanut butter cookies is completed. The copy of the peanut butter cookie recipe stored in the working PLC data file will be written over when the copy instruction copies the chocolate chip cookie recipe to the working data file recipe.

COPY INSTRUCTION PARAMETERS

The copy instruction has three parameters: source, destination, and length. *Source* is the address of the file to be copied; *Destination* is the starting address of the data file the information is to be copied to; *Length* is the number of elements you want to copy. The destination file type determines the number of words the instruction will copy. Refer to the text for additional information on the copy instruction and data types and how they affect the number of words or elements copied.

COPY INSTRUCTION PARAMETER SYNTAX

The *Copy* instruction is used to copy a block of data from one data file location to another. A user may group like data together in consecutive memory locations within a data file. As an example, a group of integer file locations containing recipe data for timing, counting, weight, mixing times, etc. would be stored together as like information pertaining to one recipe. There may be many blocks of data, each a single recipe, stored in a data file such as the integer file. Each block of data within such a file is called a user-created file.

When you want to copy a recipe with, as an example twelve recipe values, not only does the source and destination need to be specified, the source and destination must be identified as a user-created file so the CPU will know that it is to copy a consecutive block of data. To identify a block of data as a user created file, the # sign is placed before the source and the destination parameters. An example of a copy instruction is illustrated in Figure 32-1.

```
Rung 2:0
   I:1      B3                                              +COP--------------+
 --] [--- [OSR] -----------------------------------------+COPY FILE         +-
     0        2                                            |Source     #N12:20|
                                                           |Dest        #N7:10|
                                                           |Length           6|
                                                           +------------------+
```

Figure 32-1 The SLC 500 *Copy* instruction.

The *Copy* instruction illustrated specifies a source user created block of data, called a user file, starting at N12:20 (integer file twelve, element twenty) to be copied to the starting file address N7:10 (integer file seven, element ten). The length of the file is 6 elements. Since integer file elements are one word in length, six elements will be copied from the source to the destination. Remember, the destination data file type determines the number of elements / words that will be copied.

COPY INSTRUCTION STATUS BITS

Copy instruction execution does not affect any status bits.

THE LAB

This lab exercise will develop the ladder program to copy the recipe data from integer file twelve, word twenty, to integer file seven, word ten. The recipe contains six recipe parameters for time, count, weight and mixing times. For this application, assume recipe values for integer file N12:20 are as follows: 10, 20, 30, 25, 33, 12.

#N12:20			#N7:10	
N12:20	10	→	10	N7:10
N12:21	20	→	20	N7:11
N12:22	30	→	30	N7:12
N12:23	25	→	25	N7:13
N12:24	33	→	33	N7:14
N12:25	12	→	12	N7:15

Figure 32-2 Conceptual illustration of how the *Copy* instruction will operate on data during this lab exercise.

1. _____ Copy the *Begin* program from your student floppy disk to your hard drive.
2. _____ Rename the *Begin* program *Copy_1*.
3. _____ Start from the main menu screen in your APS software.
4. _____ The first thing we will do is create integer file N12, and the desired number of elements. Press F3, *Off Line Programming / Documentation*.
5. _____ Press F10, *Memory Map*.
6. _____ F6, *Create DT (Data Table) File*.

7. _____ In response to the message: "Enter Address To Create," enter *N12:99*. This will create one hundred integer file elements, 0 through 99.

8. _____ Press *Enter* to create integer file elements N12:0 through N12:99.

9. _____ Next, enter data into Integer file twelve data table.

10. _____ *Escape.*

11. _____ Save the program file.

12. _____ Press F9, *Data Monitor.*

13. _____ Enter *N12:20* to select integer file twelve, element twenty.

14. _____ Press *Enter.*

15. _____ The cursor should be blinking next to N12:20 =.

16. _____ Enter *10* and then press *Enter.*

17. _____ The value 10 should be reflected in element address N12:20 in the table.

18. _____ Press the right arrow key.

19. _____ The cursor should be blinking next to N12:21=.

20. _____ Enter *20* and press *Enter.*

21. _____ Press the right arrow key. The cursor should be blinking next to N12:22 =.

22. _____ Enter the value *30* and press *Enter.*

23. _____ Integer file addresses *N12:20, N12:21,* and *N12:22* should look similar to the partial integer file below:

	0	1	2	3	4	5	6	7	8	9
N7:20	10	20	30	0	0	0	0	0	0	0

The address is determined by starting at the left column N12:20 and observing the value across the top of the table . The first position is zero. This is N12:20. The second position is 1, which signifies element address N12:21.

24. _____ Enter the remaining numbers in file N12 as illustrated in Figure 32-2.

25. _____ When finished, press *Escape* to return to the Program Directory screen.

CREATING INTEGER FILE SEVEN'S ELEMENTS

26. _____ Even though the integer file seven has been created by default, there have been no elements created. We will create 100 elements in the integer file seven.

27. _____ To create N7 elements, press F10, *Memory Map.*

28. _____ Press F6, *Create DT File.*

29. _____ In response to the message: "Enter Address To Create", enter *N7:99*. This will create one hundred integer file elements, 0 through 99.

30. _____ Press *Enter* to create integer file elements N7:0 through N7:99. Since we will be copying data from integer file N12 to N7, we will leave the N7 file as zeros.

31. _____ When finished, press *Escape* to return to the Program Directory screen.

32. _____ Press F2, *Save.*

33. _____ F8, *Yes* to accept the defaults.

34. _____ F8, *Monitor File.*

35. _____ Press F10, *Edit.*

36. _____ Enter the ladder program from Figure 32-1 (p. 153).

37. _____ To program the copy instruction, press F7, *File Instructions.*

38. _____ Press F3 to select the copy instruction.

39. _____ Enter Source: #N12:20

40. _____ Enter Destination: #N7:10.

41. _____ Enter the length parameter as 6.

42. _____ *Enter.*

43. _____ Accept the rung.
44. _____ *Save / Go On Line.*
45. _____ Run your program.
46. _____ As the program runs, energize and then de-energize I:1/0.
47. _____ Monitor file 7.
48. _____ Fill in the table in Figure 32-3 with the data from file 7.

	0	1	2	3	4	5	6	7	8	9

Figure 32-3   Integer file seven data as a result of executing the *Copy* instruction.

QUESTIONS

1. Explain why the OSR instruction is part of the ladder diagram._____

_____

2. What would happen if I:1/0 was left energized and no OSR instruction was programmed on the ladder rung?_____

3. What keystrokes did you use to monitor data file seven?_____

4. While monitoring N7, what function key would you press to go directly to N12?_____

5. Explain the significance of the # character in the copy file instruction._____

_____

6. What are the three copy instruction parameters?_____

7. Define the first copy instruction parameter._____

_____

8. Define the second copy instruction parameter._____

_____

9. Define the third copy instruction parameter._____

_____

10. Explain the role of status bits in conjunction with the copy instruction._____

_____

33

COPY FILE
PROGRAMMING

In this exercise we will work with the copy file instruction and the integer file. We will load the integer file with numerical data and then develop program logic to copy all the previously entered words into integer file 12 locations.

THE LAB

1. _____ Refer to your text, and the SLC 500 Instruction Reference manual for more information on the Copy instruction.
2. _____ Start with the ladder you developed for exercise 32 and add the following:
3. _____ Add a program rung so when input I:1/1 is energized, the copy file instruction copies data once, starting at N7:0 and including the following twelve words, to a like file starting at N12:100.
4. _____ Create a new file N12 with enough elements to accept the copied data.

Address	0	1	2	3	4	5	6	7	8	9
N7:0										
N7:10										
N7:20										

Figure 33-1   Integer file seven's data for this exercise.

5. _____ Enter any numerical data into file N7's user-defined file, as in Figure 33-1.
6. _____ When finished editing your ladder program, go on line and run the program.
7. _____ Illustrate in Figure 33-2 (next page) where data was copied to in integer file N12.

QUESTIONS

1. What kind of data can you store in N7?_____

2. Explain what the copy instruction does._____

_____

Address:	0	1	2	3	4	5	6	7	8	9
N12:0										
N12:10										

Figure 33-2   Integer file twelve's data resulting from execution of the *Copy* instruction.

3. Illustrate how this file operates on the two integer files we will be using in this lab exercise.

_____

4. Which status bits are associated with this instruction?_____

5. Explain how status bits can be used in conjunction with this instruction._____

_____

6. Define the source parameter as it pertains to this instruction._____

_____

7. Explain when the # symbol must be used in conjunction with this instruction._____

_____

8. When the # is used, explain how it pertains to the parameters._____

_____

9. How is the length parameter used with the copy file instruction?_____

10. What is the maximum number of words that can be copied in one block?_____

11. Compose a short explanation on how this instruction and file boundaries interact._____

_____

_____

34

PROGRAMMING THE MOVE INSTRUCTION

This exercise will introduce you to the *Move* instruction and how it is used to modify rung conditions as required by changing production conditions.

ADDITIONAL MATERIAL NEEDED

A three-position selector switch connected to your PLC input section. Each switch position should have one circuit contact closure.

INTRODUCTION

Each selector switch position will close a contact block which will be wired into a PLC input screw terminal as follows:

Wire selector switch position one contact block to PLC input I:1/0.
Wire selector switch position two contact block to PLC input I:1/1.
Wire selector switch position three contact block to PLC input I:1/2.

Each selector switch position will make one of three *Move* instructions true. Each *Move* instruction will use an integer value stored in an integer file location. When a specific *Move* instruction is selected, its associated integer value will be placed in a counter's preset value. The counter preset will change in relation to the selector switch position and the value in the associated *Move* instruction. The program is illustrated in Figure 34-1 on the next page.

THE LAB

To begin developing the ladder program, start from the main APS menu.

1. _____ Copy the *Begin* program from your floppy disk and rename it *MOVE_1*.
2. _____ Press F3, *Offline Programming / Documentation*.

Rung 2:0
When the selector switch is in this position, the move instruction will move a
copy of the value from the designated source integer file element into counter
C5:0's preset value.

```
| Selector                                            When true,
| switch                                              places the
| position                                            source
| one.                                                value into
|                                                     C5:0.pre
|                                                   +MOV--------------+
|    I:1        B3                                  +MOVE           +-
----] [------[OSR]---------------------------------|Source      N7:20|
|     0          0                                  |              10|
|                                                   |Dest     C5:0.PRE|
|                                                   |              10|
|                                                   +----------------+
```

Rung 2:1
When the selector switch is in this position, the move instruction will move a
copy of the value from the designated source integer file element into counter
C5:0's preset value.

```
| Selector                                            When true,
| switch                                              places the
| position                                            source
| two.                                                value into
|                                                     C5:0.pre
|                                                   +MOV--------------+
|    I:1        B3                                  +MOVE           +-
----] [------[OSR]---------------------------------|Source      N7:21|
|     1          1                                  |              20|
|                                                   |Dest     C5:0.PRE|
|                                                   |              10|
|                                                   +----------------+
```

Rung 2:2
When the selector switch is in this position, the move instruction will move a
copy of the value from the designated source integer file element into counter
C5:0's preset value.

```
| Selector                                            When true,
| switch                                              places the
| position                                            source
| three.                                              value into
|                                                     C5:0.pre
|                                                   +MOV--------------+
|    I:1        B3                                  +MOVE           +-
----] [------[OSR]---------------------------------|Source      N7:22|
|     2          2                                  |              30|
|                                                   |Dest     C5:0.PRE|
|                                                   |              10|
|                                                   +----------------+
```

Rung 2:3
```
|    I:1                                            +CTU--------------+
--] [----------------------------------------------+COUNT UP       +-(CU)-
|     3                                             |Counter     C5:0+-(DN)
|                                                   |Preset        10|
|                                                   |Accum          0|
|                                                   +----------------+
```

Figure 34-1 *Move* instruction used in conjunction with a selector switch.

3. _____ The first thing we will do is create the elements needed in integer file seven. Even though the integer file, file seven, has been created, there have been no elements created. We will now create 100 elements in the integer file.

4. _____ Press F10, *Memory Map.*

5. _____ In response to the message: "Enter Address To Create", enter *N7:99*. This will create one hundred integer file elements, 0 through 99.
6. _____ Press *Enter* to create integer file elements N7:0 through N7:99.
7. _____ *Escape* to return to the program directory screen.
8. _____ F2, *Save*.
9. _____ Press *Yes* to accept the defaults.
10. _____ F8, *Monitor File*.
11. _____ F10, *Edit*.

ENTERING THE LADDER PROGRAM FROM FIGURE 34-1

12. _____ Program the XIC instruction and OSR instruction on rung zero.
13. _____ To program the move instruction, press F6, *Move / Logical*.
14. _____ Press F1, *MOV* to select the move instruction.
15. _____ Enter Source: *N7:20*.
16. _____ Enter Destination: *C5:0.PRE*.
17. _____ Accept the rung.

    This *Move* instruction will move a copy of data stored in source N7:20 to the destination C5:0.Preset. We will enter data into integer file locations after we complete developing our ladder rungs.

18. _____ Program the remaining rungs on your own.
19. _____ After accepting the last program ladder rung, press *Escape*.

ENTERING DATA INTO THE INTEGER FILE SEVEN DATA TABLE

20. _____ Press F8, *Data Monitor*.
21. _____ Enter N7:20 to select integer file seven, element twenty.
22. _____ Press *Enter*.
23. _____ The cursor should be blinking next to N7:20.
24. _____ Enter *10* and press *Enter*.
25. _____ The value 10 should be reflected in element address N7:20.
26. _____ Press the right arrow key.
27. _____ The cursor should be blinking next to N7:21=.
28. _____ Enter *20* and press *Enter*.
29. _____ Press the right arrow key. The cursor should be blinking next to N7:22=
30. _____ Enter the value *30* and press *Enter*.
31. _____ Integer file addresses N7:20, N7:21, and N7:22 should look like the partial integer file below:

	0	1	2	3	4	5	6	7	8	9
N7:20	10	20	30	0	0	0	0	0	0	0

    The address is determined by starting at the left column, N7:20, and observing the value across the top of the table . The first position is zero. This is N7:20. The second position is 1, this signifies element address N7:21.

32. _____ *Escape* to return to your ladder program.
33. _____ Notice that each move instruction source parameter reflects the data stored at that address.
34. _____ *Save / Go On Line*.
35. _____ Run your program.

36. _____ As the program runs, move the selector switch to different positions. Observe the count preset values change as the commanded values are moved into it.

QUESTIONS

Answer the questions below based on data in the sample integer data file in Figure 34-2.

Address	0	1	2	3	4	5	6	7	8	9
N7:0	1	2	3	4	5	6	7	8	9	10
N7:10	100	110	120	130	5	10	15	20	25	30
N7:20	10	20	30	0	0	4	6	9	1	4
N7:30	2	4	6	8	10	12	14	16	18	20
N7:40	20	25	30	35	40	45	52	63	74	89

Figure 34-2

1. N7:10 = _____
2. N7:25 = _____
3. N7:42 = _____
4. N7:2 = _____

5. N7:9 = _____
6. N7:0 = _____
7. N7:33 = _____
8. N7:46 = _____

35

UP-COUNTERS AND DOWN-COUNTERS

This exercise will give you practice programming up- and down-counters in a simulated application.

THE LAB, PART I

1. _____ Copy *Timer_3* program from your floppy disk to your personal computer's hard drive.
2. _____ Rename the *Timer_3* program *Up_Dwn*.
3. _____ From the monitor file screen, press F10, *Edit*.
4. _____ Cursor to Rung 2:1
5. _____ Press *Delete Rung*, F6
6. _____ Delete the new rung 2:1. This should be the rung with the T4:0/EN instruction.
7. _____ Press F6, *Delete Rung*.
8. _____ With these two rungs deleted, Rung 2:1 will now be the rung T4:0/DN controlling output 0:2/1.
9. _____ We are going to edit this rung and replace the output instruction with a counter.
10. _____ Cursor to the 0:2/1 instruction.
11. _____ Press F5, *Modify Rung*.
12. _____ Press F5, *Modify Instruction*.
13. _____ Press F2, *Timer/Counter*.
14. _____ Press F4, *CTU* (Count Up Counter).
         Fill in counter parameters:
           Enter counter address >*C5:0*
           Preset > *10*
           Enter accumulated value > *0*
15. _____ Press *Escape*.
16. _____ Press F10, *Accept Rung*.

Adding Documentation

17. _____ Press *Escape*.

18. _____ Press F5, *Documentation.*
19. _____ Press F1, *Rung Comment.*
      Enter documentation text as illustrated in Figure 35-1.

```
Rung 2:0
The T4:0/DN XIO instruction makes this a free running timer.  As a free
running timer, the timer done bit will be energized every thirty seconds. The
done bit will energize for one processor scan. Thirty seconds is the preset
value. Each time the preset is equal to the accumulated value, the done bit
will be made true. Being true, the done bit will remove power from the TON
instruction for one scan, causing the instruction to reset to zero.
|  I:1   T4:0                                            +TON-------------+     |
|--] [---]/[---------------------------------------------+TIMER ON DELAY   +-(EN)-|
|   0    DN                                              |Timer        T4:0+-(DN) |
                                                         |Time Base     1.0|      |
                                                         |Preset         30|      |
                                                         |Accum           0|      |
                                                         +-----------------+      |
```

```
Rung 2:1
Count up counter will increment each time T4:0's done bit is pulsed.  Since
T4:0 is a free-running timer with a pulse to the done bit every 30 seconds,
C5:0 will increment every 30 seconds.
|  T4:0                                                  +CTU-------------+      |
|--] [--------------------------------------------------+COUNT UP         +-(CU)-|
|   DN                                                  |Counter      C5:0+-(DN) |
                                                        |Preset         10|      |
                                                        |Accum           0|      |
                                                        +-----------------+      |
```

```
Rung 2:2
|  C5:0                                                                    O:2  |
|--] [-------------------------------------------------------------------( )--  |
|   DN                                                                      2   |
```

```
Rung 2:3
|  I:1                                                                     C5:0 |
|--] [------------------------------------------------------------------(RES)-- |
|   3                                                                          |
```

Figure 35-1 *Up_Down* counter program, part 1.

20. _____ F8, *Accept / Exit.*
21. _____ F10, *Save Documentation.*
22. _____ *Escape.*
23. _____ F10, *Edit.*
24. _____ With Cursor on the rung you just edited, Press F3, *Append Rung.*
25. _____ F4, *Insert Instruction.*
26. _____ Press F1, *Bit.*
27. _____ Press F1, *XIC.*
28. _____ Enter bit address > *C 5:0/DN.*
29. _____ *Enter.*
30. _____ Press F1, *Bit.*
31. _____ Press F3, *OTE.*
32. _____ Enter Bit Address> *0:2/2.*
33. _____ *Escape.*
34. _____ Accept rung.
35. _____ To automatically append a new rung, press F4, *Insert Instruction.*
36. _____ Press F1, *Bit.*

37. _____ Press F1, *XIC*.
38. _____ Enter bit Address > *I:1/3*.
39. _____ *Enter*.
40. _____ Press F2, *Timer / Counter*.
41. _____ Press F8, *Reset*.
42. _____ Enter Structure Address > *C5:0*.
43. _____ *Enter*.
44. _____ *Escape*.
45. _____ Accept rung.
46. _____ Press *Escape* to remove extra appended rung.
47. _____ Add documentation to all rungs explaining their function in the program.
48. _____ Press F10, *Edit*.
49. _____ Press F1, *Save and Go Online*.
50. _____ Put processor in run mode.

Questions

1. Identify the comment above the T4:0/DN instruction. What type of comment is this?
   _____ Instruction
   _____ Rung
   _____ Address

2. From the Edit Screen: what six key strokes do you enter to type in the Comment from question one? _____

PART II: COUNTER PROGRAMMING PRACTICE

Assume our timer T4:0, and T4:0 DN is actually representing a signal coming in from a conveyor line. Every thirty seconds one filled bottle passes the photo-electric sensor. This sensor input is represented by the T4:0 DN bit. Counter C5:0 will track how many bottles have been produced.

For this application, develop the PLC program, after which your partner will install an operator interface display device that will show the machine operator the remaining bottles that need to be filled to complete the production run. Counter C5:0 will track how many bottles have been produced. To track how many bottles remain to be filled, you need to add a count-down counter (C5:1) to the current program. Use T4:0/DN bit to trigger the CTD instruction to decrement. When the down counter, C5:1 reaches zero, it automatically resets to the batch total value. C5:0 is reset by the operator after recording the shift to total of bottles produced.

The Lab

1. _____ Edit the *Up_Down* program.
2. _____ Change the preset for C5:0 to 10,000.
3. _____ Insert a new rung between rungs one and two.
4. _____ Program the new rungs as illustrated in figure 35-2 on the next page.
5. _____ Set CTD counter preset to 0.
6. _____ Set CTD counter accumulated value to 100.
7. _____ Accept the rung.
8. _____ Add rung comment as illustrated.
9. _____ *Escape*.
10. _____ Data monitor.
11. _____ Press F5, *Specify Address*.
12. _____ Enter data table address *N* (Integer file 7).

Rung 2:0
The T4:0/DN XIO instruction makes this a free running timer. As a free
running timer, the timer done bit will be energized every thirty seconds. The
done bit will energize for one processor scan. Thirty seconds is the preset
value. Each time the preset is equal to the accumulated value, the done bit
will be made true. Being true, the done bit will remove power from the TON
instruction for one scan, causing the instruction to reset to zero.

```
 I:1    T4:0                                    +TON--------------+
--] [---]/[------------------------------------+TIMER ON DELAY    +-(EN)-
  0     DN                                      |Timer        T4:0+-(DN)
                                                |Time Base     1.0|
                                                |Preset         30|
                                                |Accum           0|
                                                +-----------------+
```

Rung 2:1
Count up counter will increment each time T4:0's done bit is pulsed. Since
T4:0 is a free-running timer with a pulse to the done bit every 30 seconds,
C5:0 will increment every 30 seconds. The value 10,000 was arbitrarily
selected. 10,000 bottles would be equal to 100 batches, this is more than a
shift produces each day. This will be our counter for tracking total
production.

```
 T4:0                                           +CTU--------------+
--] [------------------------------------------+COUNT UP          +-(CU)-
  DN                                            |Counter      C5:0+-(DN)
                                                |Preset      10000|
                                                |Accum           0|
                                                +-----------------+
```

Rung 2:2
The accumulated value must be set to the beginning value the counter is to
count down from. Each time the input instruction transitions from false to
true, the counter will decrement the accumulated value by one.

```
 T4:0                                           +CTD--------------+
--] [------------------------------------------+COUNT DOWN        +-(CD)-
  DN                                            |Counter      C5:1+-(DN)
                                                |Preset          0|
                                                |Accum         100|
                                                +-----------------+
```

Rung 2:3
```
 C5:0                                                          O:2
--] [------------------------------------------------------- ( )--
  DN                                                            2
```

Rung 2:4
```
 I:1                                                          C5:0
--] [------------------------------------------------------ (RES)--
  3
```

Rung 2:5
The move instruction is used to move a new beginning batch total back into the
CTD counter.
```
 C5:1                                           +MOV--------------+
--] [------------------------------------------+MOVE             +-
  DN                                            |Source      N7:14|
                                                |             100|
                                                |Dest     C5:1.ACC|
                                                |             100|
                                                +-----------------+
```

Figure 35-2 *Up-Down* counter programming practice program.

13. _____ *Enter.*
14. _____ The message, " No data in integer file 7," means that, though the file was created as a
default data file, no elements were created.
15. _____ *Escape.*
16. _____ F3, *Exit.*
17. _____ F10, *Memory Map.*

18. _____ Press F6 to create elements in integer file 7. Notice file 7 contains 0 elements and 0 words.
19. _____ Enter the address to create: N7:99 to create 100 elements in the integer file.
20. _____ Enter.
21. _____ Notice that integer file 7 now contains 100 elements and 100 words.
22. _____ Escape.
23. _____ F2, Save.
24. _____ F8 to accept defaults.
25. _____ F9, Data Monitor.
26. _____ Enter data table address N7.
27. _____ Enter.
28. _____ Move cursor until Integer file address N7:14 is highlighted.
29. _____ The cursor should be blinking next to N7:14=.
30. _____ Enter 100 and press Enter.
31. _____ Escape.
32. _____ F2, Save.
33. _____ F8 to accept defaults.
34. _____ Add a rung that will move the counter accumulated value of 100 from N7:14 into C5:1.ACC.
35. _____ Run program and observe counter and timer operation.

QUESTIONS

1. Why was the *Move* instruction necessary?_____

2. What is the largest preset the counter will accept?_____

3. When the accumulated value is less than the preset value, what happens to the accumulated value if the counter continues to count down?_____

_____

5. Are any status bits affected? Which ones? Explain what happens._____

_____

6. If counter continues to count down, explain how the counting will decrement._____

_____

SUMMARY

For this exercise we developed an up-down counter, controlled by the same input, tracking parts produced. The count-up counter tracks the running total of parts produced. The down-counter will count each batch of 100 parts. When the down counter reaches zero, a *Move* instruction is necessary to move the accumulated value 100 back into the CTU ACCUM parameter.

36

UNDERSTANDING HEXADECIMAL MASKS

Before working with the sequencer instruction, we need to understand the workings of hexadecimal masks. Since the surest way to learn is by doing, we will experiment with the *masked move instruction*. The masked move instruction is an easy way to understand how a Hex mask works. To understand how the hex mask works we will take data from the integer file N7, move this data through a hex mask, and view the result on the LEDs on an output module.

Before we look at the masked move instruction, let's review the basic *move* instruction. The move instruction moves a *copy* of the data contained in the source location to the designated destination. So the term *move instruction* is a bit deceiving, as it implies moving data *from* one location to another. In fact, the move instruction places a copy of the data in the designated destination, leaving the original data in place at the source location.

The move instruction is an output instruction with two parameters:

A. The *source* is an address where data is stored, or a constant entered into the source parameter of the instruction.

B. The *destination* is the address where a copy of the data is to be placed, or "moved."

If a copy of the data is to be placed only once, a one-shot instruction should be programmed ahead of the move instruction. Figure 36-1 on the next page illustrates a move instruction used in conjunction with a one-shot.

Evaluating the rung, each time I:1/2 becomes true, the one-shot will trigger the move instruction. The move instruction (*MOV*) will place a copy of the data contained in N7:0 in N:15. The data *123* which is contained in N7:0 will also be found in N7:15.

THE LAB, PART I

First we will create words of data in integer file N7. The move instruction will then be programmed to move the data from one location to another.

1. _____ Load the *Begin* program from your floppy disk.
2. _____ Rename the processor file to *LAB_MOV*.
3. _____ Go to the Monitor file screen.

```
Rung 2:0
When XIC I:1/2 is true, the One-shot will allow the move instruction to
execute one time. A copy of the data, 123 contained in N7:0 will placed in
destination, N7:15.  The original data will still be stored in N7:0.
   I:1      B3                                              +MOV--------------+
 --] [--- [OSR] -----------------------------------------+MOVE              +-
     2        1                                           |Source      N7:0|
                                                          |             123|
                                                          |Dest       N7:15|
                                                          |               0|
                                                          +-----------------+
```

Rung 2:1

Figure 36-1 *Move* instruction used in conjunction with a one-shot instruction.

4. _____ Press F7, *General Utility*.

5. _____ F1, *Memory Map*.

6. _____ On the data table map screen, under the column, elements and words, notice that the integer file has a zero listed (as do some others). This means there are zero elements existing in these files. Before we can store data in the integer file, we will have to create data elements for the information to be stored in. To add elements to integer file 7:

7. _____ Press F6, *Create Data Table File*. Enter the address to create: *N7:49*. We will create 50 elements, 0-49.

8. _____ Press *Enter*. In Figure 36-2 the memory map will now show that N7 contains fifty words, addresses from N7:0 through N7:49. Remember, the integer file has one-word elements, thus the element count and word count will be identical.

```
                         DATA TABLE MAP
FILE        TYPE         LAST ADDRESS     ELEMENTS WORDS
  0    0 output          0:6                    4     4    FILE PROTECTION
  1    I input           I:1                    1     1    STATIC
  2    S status          S:32                  33    33
  3    B binary or bit   B3/15                  1     1
  4    T timer           T4:0                   1     3
  5    C counter                                0     0
  6    R control                                0     0
  7    N integer         N7:49                 50    50
  8      reserved                               0     0


                    PROCESSOR MEMORY LAYOUT
      92 data words of memory used in   9 data table files
     230 instruction words of memory used in   3 program files
    3843 instruction words of unused memory available


Press a key or enter file number

offline             SLC 5/02
                                                          File
                              CREATE  DELETE
                              DT FILE DT FILE                   DATA
                                F6      F7                     PROTECT
                                                                F10
```

Figure 36-2 Data table map.

9. _____ *Escape*.

10. _____ *Escape* to return to the screen with your end program rung.

Programming the Move Instruction

1. _____ To begin developing the ladder rung from figure 36-1, press F10, *Edit*.
2. _____ F4, *Insert Rung*.
3. _____ Insert instruction.
4. _____ F1, *Bit*.
     Enter bit address> *I:1/2*.
5. _____ *Enter*.
6. _____ F1, *Bit*.
7. _____ F6, *OSR* (One-shot instruction).
     Enter bit address> *B3:0/1*.
8. _____ *Enter*.
9. _____ Select F6, for the group of *Move / Logical instructions*.
10. _____ F1, *Move*.
     Enter source address or value> *N7:0*.
     Enter.
     Enter destination address > *N7:15*.
     Enter.
11. _____ *Escape*.
12. _____ Accept rung.

Entering Data into the N7 Integer File

1. _____ *Escape*.
2. _____ Move the cursor to highlight the *Move* instruction.
3. _____ F8, *Data Monitor*.
4. _____ Use the arrow keys to move cursor to N7:0.
5. _____ Enter the number *123*.
6. _____ *Enter*.
7. _____ Using the right arrow, move the cursor to N7:1.
8. _____ Enter the number *10922*.
9. _____ *Enter*.
10. _____ Enter the number *28588* into N7:2.
11. _____ *Escape*.
12. _____ Save and go on-line.
13. _____ Run your program.
14. _____ While the program runs, energize I:1/2.
15. _____ Monitor Data.
16. _____ Monitor N7:15. A copy of the value *123* should have been moved from N7:0 to N7:15. The original data will still be in location N7:0.
17. _____ Escape to return to your ladder program.

Masked Move Instruction

In some instances we may not desire to move all sixteen source bits to a specific destination. If only the lower eight bits of the source word to the destination, we need a way to block, or mask out, the bits we do not desire to move. This can be accomplished by passing the bits through a controlling mechanism which allows us to control which bits pass through to the destination. This controlling mechanism is called a *mask*.

The *masked move instruction* moves a *copy* of the data contained in the source location through a *hexadecimal mask* to the designated destination. A copy of the original data is moved through a user-designated hexadecimal mask parameter to the designated destination; the original data remains in the source location.

The masked move instruction is an output instruction with three parameters:

A. The *source* is an address where data is stored, or a constant entered into the source parameter of the instruction.

B. The *mask* is the address where the mask will be found, or a constant hex value.

C. The *destination* is the address where a copy of the data is to be placed, or moved.

The masked move instruction will move data to the designated destination each scan the instruction is true. If a copy of the data is to be moved only once, a one-shot instruction should be programmed ahead of the move instruction. In Figure 36-3, rung 2:1 illustrates a masked move instruction used in conjunction with a one-shot.

```
Rung 2:1
When the input is true the OSR will trigger the MVM intstruction once. Data
contained in N7:1 will be passed through the mask  to the destination O:2.0.
   I:1      B3                                                +MVM-------------+
 --] [--- [OSR] -----------------------------------------+MASKED MOVE      +-
     3       2                                            | Source        N7:1|
                                                          |               10922|
                                                          | Mask          00FF|
                                                          |
                                                          | Dest          O:2.0|
                                                          |                  0|
                                                          +-----------------+
```

Figure 36-3 The *Masked Move* instruction.

Evaluating the rung, each time I:1/3 becomes true, the one shot will trigger the masked move instruction. The masked move instruction (*MVM*) will move a copy of the data contained in N7:1 to N:12 through the hex value designated as the mask .

Applying Hexadecimal Numbers and Masks with Slc 500 Programming Instructions

The SLC 500 mask parameter uses either a four-character hexadecimal numerical code as the mask word, or the address where the mask word will be found. This hex word, used in its binary equivalent form, is the mask through which the masked move instruction will move the data. If the mask bits are *set,* or a logical one, data will pass through the mask, while bits that are *reset,* a logical zero, the mask will restrict data from passing. Let's look at an example.

EXAMPLE ONE

Figure 36-4 illustrates a source word 0110 1111 1010 1100. The lower eight bits need to pass through the mask to the destination. The upper eight bits of the source word are to be blocked by the mask. As a result, the lower eight source bits will be reflected in the destination word, while the upper eight destination bits will remain unchanged. Figure 36-4 identifies bits to pass with a Y for yes and an N for no.

0110	1111	1010	1100	Source word
NNNN	NNNN	YYYY	YYYY	Bit to pass through to destination?
Bits do not change		1010	1100	Destination word

Figure 36-4 The lower eight source bits will be allowed to pass through mask.

To set up the mask to pass the correct data, set a 1 in the mask bit position where each source data bit is to pass. This is illustrated in Figure 36-5.

0110	1111	1010	1100	Source word
0000	0000	1111	1111	Bit to pass through to destination?
Bits do not change		1010	1100	Destination word

Figure 36-5   Bits allowed to pass move through 1's in Hex mask.

Next, convert bit format to Hex, as illustrated in Figure 36-6. The converted hex value will be entered into the mask parameter of the masked move instruction.

The mask bit pattern:	0000	0000	1111	1111
Hex value of:	0	0	F	F

Figure 36-6   Mask bits translated to HEX.

EXAMPLE TWO

The next example will illustrate how to pass only selected bits in the lower byte of the source data word to the destination. Note that the X's in the destination word mean these bits do not change from their original contents.

0110	1111	1010	1100	Source word
NNNN	NNNN	YYNN	NNYY	Bit to pass through to destination?
Bits do not change		10XX	XX00	Destination word

Figure 36-7   Y's indicate which bits are to pass through the mask.

To set up the mask so as to pass the correct data, you simply set a *1* in the mask bit position where each source data bit is to pass. This is illustrated in Figure 36-8.

0110	1111	1010	1100	Source word
0000	0000	1100	0011	Bit to pass through to destination?
Bits do not change		10XX	XX00	Destination word

Figure 36-8   Y's from Figure 36-7 changed to 1's.

Convert bit format to Hex; this will be entered into the mask parameter of the masked move instruction, as illustrated in Figure 36-9.

The mask bit pattern:	0000	0000	1100	0011
Hex value of:	0	0	C	3

Figure 36-9   Mask binary bit pattern converted to hex.

Mask Data Representation

Hexadecimal numbers are base sixteen. Any valid hex number from 0 to F can be used to select the combination of four bits which will be used in mask construction. Figure 36-10 lists hexadecimal numbers and their corresponding bit patterns.

HEXADECIMAL CORRESPONDING BIT PATTERNS			
0	0000	8	1000
1	0001	9	1001
2	0010	A	1010
3	0011	B	1011
4	0100	C	1100
5	0101	D	1101
6	0110	E	1110
7	0111	F	1111

Figure 36-10 Hex and binary bit pattern correlations.

THE LAB, PART II: DEVELOPING A MASKED MOVE LADDER PROGRAM RUNG

We will now develop a ladder rung for each of previous examples. Our first example of the masked move instruction started with the source word of 0110 1111 1010 1100 and moved it through a 00FF mask. Figure 36-11 illustrates the program rung we will be adding to our *Lab_Mov* program.

```
Rung 2:0
When XIC I:1/2 is true, the One-shot will allow the move instruction to
execute one time. A copy of the data, 123 contained in N7:0 will placed in
destination, N7:15.  The original data will still be stored in N7:0.
  I:1      B3                                                  +MOV--------------+
--] [---[OSR]-------------------------------------------------+MOVE             +-
   2        1                                                 |Source       N7:0|
                                                              |              123|
                                                              |Dest        N7:15|
                                                              |                0|
                                                              +-----------------+

Rung 2:1
When the input is true the OSR will trigger the MVM intstruction once. Data
contained in N7:1 will be passed through the mask  to the destination O:2.0.
  I:1      B3                                                  +MVM--------------+
--] [---[OSR]-------------------------------------------------+MASKED MOVE      +-
   3        2                                                 |Source       N7:1|
                                                              |            10922|
                                                              |Mask         00FF|
                                                              |                 |
                                                              |Dest        O:2.0|
                                                              |                0|
                                                              +-----------------+
```

Figure 36-11 Edited *Lab_Mov* ladder program.

We will be adding logic to the current ladder program.

1. _____ Start with *Lab_Mov* Edit screen.
2. _____ Move cursor to rung zero.
3. _____ Append rung.
4. _____ Insert instruction.
5. _____ F1, *Bit*.
6. _____ F1, *XIC*.
           Enter bit address> *I:1/3*
7. _____ *Enter.*

8. _____ Press F1, *Bit.*
9. _____ F6, *OSR.*
            Enter bit address> *B3:0/2*
10. _____ *Enter.*
11. _____ F6, *Move / Logical,* to select the masked move instruction.
            Enter source address> *N7:1*
            Enter.
            Enter mask address or value> *00FF*
            Enter.
            Enter destination address> *O:2.0*
            Enter.
12. _____ *Escape.*
13. _____ *Accept Rung.*
14. _____ *Save / Go on line.*

Testing Your Program

Fill in the table in Figure 36-12 with the source and mask information. Notice that the MVM source parameter is the decimal number 10922 and the mask is in Hex. To view the data file and see the binary equivalent of the decimal value in N7:1:

1. _____ Press F8, *Data Monitor.*
            Enter the data table address N7:1
            Enter.
2. _____ F1, *Change Radix.*
3. _____ F1, *Binary.*
4. _____ Record the binary bit pattern in the source area of Figure 36-12.
5. _____ Record the mask in the mask area of Figure 36-12.

				Source word
				Mask
				Expected destination word

Figure 36-12 Expected result of executing rung 2:1.

6. _____ Fill in the expected destination word in Figure 36-12.
7. _____ Energize input I:1/3. The MVM instruction will be executed once.
8. _____ What output LED's on the output module in slot two are energized?
9. _____ Does the output bit pattern correspond to the expected values you entered in Figure 36-12?

Viewing the Output Status Table Bit Pattern

Let's look at the output status table bit pattern and compare it to the destination word from figure 36-12.

1. _____ F8, *Data Monitor.*
2. _____ Enter data table address *O* to signify the output status file.
3. _____ Does the bit pattern for O:2 confirm your observations?
4. _____ *Escape* to return to your ladder program.

THE LAB, PART III

We will now develop a ladder rung for rung two of the previous example. This example of the masked move instruction source word is 0110 1111 1010 1100 and is moved through a 00C3 mask. Figure 36-13 illustrates the program rung we will be adding to our *Lab_Mov* program.

```
Rung 2:0
When XIC I:1/2 is true, the One-shot will allow the move instruction to
execute one time. A copy of the data, 123 contained in N7:0 will placed in
destination, N7:15.  The original data will still be stored in N7:0.
    I:1     B3                                              +MOV--------------+
 --] [--- [OSR]----------------------------------------+MOVE             +-
      2       1                                         | Source       N7:0|
                                                        |               123|
                                                        | Dest        N7:15|
                                                        |                  0|
                                                        +------------------+

Rung 2:1
When the input is true the OSR will trigger the MVM instruction once. Data
contained in N7:1 will be passed through the mask to the destination O:2.0.
    I:1     B3                                              +MVM--------------+
 --] [--- [OSR]----------------------------------------+MASKED MOVE      +-
      3       2                                         | Source       N7:1|
                                                        |             10922|
                                                        | Mask         00FF|
                                                        |                  |
                                                        | Dest        O:2.0|
                                                        |                  0|
                                                        +------------------+

Rung 2:2
When the input is true, the OSR will trigger the MVM instruction once.  Data
contained in N7:1 will be passed through the mask to the destination, O:3.0

    I:1     B3                                              +MVM--------------+
 --] [--- [OSR]----------------------------------------+MASKED MOVE      +-
      4       3                                         | Source       N7:2|
                                                        |             28588|
                                                        | Mask         00C3|
                                                        |                  |
                                                        | Dest        O:3.0|
                                                        |                  0|
                                                        +------------------+
```

Figure 36-13   Edited *Lab_Mov* ladder program containing new rung 2:2.

We will be adding logic illustrated in Figure 36-13 to the current ladder program.

1. _____ Program rung two as illustrated in Figure 36-13, using the same basic procedure you used to program rung one.
2. _____ Record the binary bit pattern of the source in Figure 36-14.
3. _____ Record the mask in Figure 36-14.
4. _____ Fill in the expected destination word in Figure 36-14.
5. _____ Energize input I:1/4. The MVM instruction will be executed once.

				Source word
				Mask
				Expected destination word

Figure 36-14   Expected result of executing rung 2:2.

6. _____ What output LED's on the output module in slot three are energized?
7. _____ Does the output bit pattern correspond to the expected values you entered in figure 36-14?
8. _____ Look at the output status table bit pattern and compare it to the destination word from Figure 36-12.

Changing Source Data for the Mvm on Rung Two

This exercise will edit the MVM instruction on rung two and change the source address from N7:1 to N7:2.

1. _____ From the edit ladder program screen, press F5, *Modify Rung*.
2. _____ Position the cursor so it is on the MVM instruction.
3. _____ F5, *Modify Instruction*.
> Press *Enter* to edit the MVM instruction.
> Enter the new source address: *N7:2*.
> Press *Enter*.
> Leave mask as 00C3.
> *Enter*.
> Leave destination as O:3.0
> Press *Enter*.
4. _____ *Escape*.
5. _____ F10, *Accept Rung*.
6. _____ Record the binary bit pattern of the source in Figure 36-15.
7. _____ Record the mask in Figure 36-15.
8. _____ Fill in the expected destination word in Figure 36-15.
9. _____ *Save / Go on line*.
10. _____ Energize I:1/4. Do the PLC output module LEDs correspond to the expected result?

				Source word
				Mask
				Expected destination word

Figure 36-15   Expected result of executing modified rung 2:2.

Changing Masks

This exercise will edit the MVM instruction on rung two and change the mask. Use the same procedure you used to edit your ladder program rung to change the source address; but, rather than changing the source address, change the mask as specified below. As your program runs observe the output module's LEDs and verify that the output data word corresponds to the expected destination word for the MVM instruction.

Practice I

1. _____ Edit rung two so the mask is now ABCD. Complete editing, save and go on line, verify the observed results.
2. _____ Record the binary bit pattern of the source in Figure 36-16 (next page).
3. _____ Record the mask in Figure 36-16.
4. _____ Fill in the expected destination word in Figure 36-16.
5. _____ *Save / Go on line*.
6. _____ Energize I:1/4. Do the PLC output module LEDs correspond to the expected result?

				Source word
				Mask
				Expected destination word

Figure 36-16 Expected result of executed modified rung 2:2.

Practice II

1. _____ Edit rung two so the mask is now 0F0F. Complete editing, save and go on line, verify the observed results.
2. _____ Record the binary bit pattern of the source in Figure 36-17.
3. _____ Record the mask in Figure 36-17.
4. _____ Fill in the expected destination word in Figure 36-17.
5. _____ *Save / Go on line.*
6. _____ Energize I:1/4. Do the PLC output module LED's correspond to the expected result?

				Source word
				Mask
				Expected destination word

Figure 36-17 Expected result of executing modified rung 2:2.

Practice III

1. _____ Edit rung two so the mask is now FF00. Complete editing, save and go on line, verify the observed results.
2. _____ Record the binary bit pattern of the source in Figure 36-18.
3. _____ Record the mask in Figure 36-18.
4. _____ Fill in the expected destination word in Figure 36-18.
5. _____ Save / Go on line.
6. _____ Energize I:1/4. Does the PLC output module LED's correspond to the expected result?

				Source word
				Mask
				Expected destination word

Figure 36-18 Expected result of executing modified rung 2:2.

Practice IV

1. _____ Edit rung two so the mask is now address N7:3.
2. _____ Monitor the data table N7. Enter DCBA as mask data in N7:3 Complete editing, save and go on line, verify the observed results.
2. _____ Record the binary bit pattern of the source in Figure 36-19.
3. _____ Record the mask in Figure 36-19.
4. _____ Fill in the expected destination word in Figure 36-19.
5. _____ *Save / Go on line.*
6. _____ Energize I:1/4. Do the PLC output module LEDs correspond to the expected result?

				Source word
				Mask
				Expected destination word

Figure 36-19 Expected result of executing modified rung 2:2.

EXERCISES

1. Fill in the destination word in Figure 36-20. Assume the beginning destination word is 0000 0000 0000 0000 unless instructed otherwise. There may be more then one correct answer.

1010	1010	1010	1010	Source word
0000	1111	0000	1111	Mask word
				Destination word

Figure 36-20

The hex mask is _____.

2. Fill in the mask and destination word in Figure 36-21 if the hex mask is 00FF.

1110	1001	0011	0001	Source word
				Mask word
				Destination word

Figure 36-21

3. Fill in the source word in Figure 36-22 if the hex mask is ABCD.

				Source word
A	B	C	D	Mask word
1000	0011	0010	1101	Destination word

Figure 36-22

4. Fill in hex mask in Figure 36-23 if:

1111	1000	1010	1101	Source word
				Mask word
1101	0000	0000	1000	Destination word

Figure 36-23

5. Fill in mask and destination in Figure 36-24 if we wanted to mask out the upper byte.

1101	0010	0000	0111	Source word
				Mask word
				Destination word

Figure 36-24

6. Fill in mask and destination in Figure 36-25 if we wanted to only allow bits 6 through 13 to pass.

1110	0100	1100	0001	Source word
				Mask word
				Destination word

Figure 36-25

7. Your source word is 1101111011010111. The mask is F0B4. What is the destination word?

8. Your destination word is 1011011011011010. The mask is 1C2B. What is the source word?

37

SEQUENCERS

The *sequencer* instruction has become one of the workhorse features of the PLC. The sequencer simply controls a predetermined sequence of events, for example the control of a pallet stretch wrap machine. Each step of the pallet wrapping routine is controlled by preprogrammed sequencer steps entered into the PLC user program. Each step of the sequence performs a predetermined task, such as:

1. Determine if adequate stretch wrap file is available.
2. Sense presence of a pallet to be wrapped.
3. Move pallet to proper position for wrapping.
4. Perform the wrapping process.
5. PLC is programmed to remember stretch wrap patterns. Plastic wrap is staggered to improve transportability of the loaded pallet. Each staggered wrapping pass is a sequence in itself.
6. When the wrapping process is completed, pallet is sequenced to exit the conveyor section to be loaded into a truck, or possibly an automatic guided vehicle for transport to an indexed storage location.

Instead of your user program solving your logic and sending the solved logical status to the output status table and then to the output module when outputs are updated, the sequencer instruction sends a predetermined 16-bit data word representing the desired output configuration for a specific output module at a sequence step in a timed relationship to the output status file. The output status file turns on and off the outputs in the proper sequence during the output update portion of the scan. The SLC 500 sequencer output instruction, SQO, steps through the sequencer file, usually a bit file pattern whose bits have been previously set up to control the desired output devices.

THE LAB

In this exercise, we will develop a sequencer data table in a bit file where each 16-bit word in the data file will represent one step in our sequence. We will use a timer to step from one position (or "step") in the sequence to the next, to control (or "sequence") our outputs on and off.

DEVELOPING A SEQUENCER LADDER PROGRAM

We will develop the following ladder rung (Figure 37-1) containing an input and the SQO instruction:

```
Rung 2:0
Each time I:1/0 is cycled, the sequencer instruction (SQO) will increment to
the next position, or step in the sequence.  When all of the steps have been
stepped through (the length) the sequence will begin at the first step and
step through the sequence again.
   I:1                                                  +SQO--------------+
 --] [-------------------------------------------------+SEQUENCER OUTPUT  +-(EN)-
      0                                                |File       #B12:0+-(DN)
                                                       |Mask         FFFF|
                                                       |Dest        O:2.0|
                                                       |Control      R6:1|
                                                       |Length         16|
                                                       |Position        0|
                                                       +----------------+
```

Figure 37-1 *SEQ_ONE program.*

1. _____ Load your *Begin* program from your floppy disk.
2. _____ Rename the file *SEQ_ONE*.
3. _____ *Monitor File*.
4. _____ *Edit*.
5. _____ *Insert Rung*.
6. _____ Insert *XIC* Instruction with address I:1/0.
7. _____ To enter sequencer instruction select F8.
8. _____ Select F5, *SQO*.
9. _____ Enter File: #*B:12:0* (We will build new data file B:12).
10. _____ Enter Mask: *FFFF*.
11. _____ Enter Destination:*O:2* (Output module slot 2)
12. _____ Enter Control Register Structure Address > *R6:1*.
13. _____ Enter Length: *16* (number of steps)
14. _____ Enter Position: *0*.
15. _____ *Accept the Rung*.

BUILD THE SEQUENCER'S DATA TABLE

To create a new bit file, B:12.
1. _____ From Main Menu select F3.
2. _____ Rename the file *SEQ_ONE*.
3. _____ *Monitor File*
4. _____ *Edit*.
5. _____ F6, *Create Data File*.
6. _____ Enter the address *B12:24* to create bit file 12 with twenty five 16-bit words.
7. _____ *Escape*.

DATA TABLE CONFIGURATION

Now that B:12 has been created, we need to add the proper sequencer step data to each bit file word, or element.

For this exercise, our sequencer is to step through each output on the module in slot two (O:2.0). The first step of the sequence will start at output zero and move one output at a time until each output on the 16-point output module or fixed output points have been turned on and off in a sequential fashion. Figure 37-2 illustrates a screen print of the completed data table file B12. Each sequence step has been identified to the left of each bit file word (this is for illustration only and does not appear in the actual data file). Notice there is a binary 1 in each data file word (representing a sequencer step) representing the output point to be energized during each sequencer step.

```
            address        15      data       0              address        15      data       0
Step 0 ->B12:0         0000 0000 0000 0000  Step 16->B12:16    1000 0000 0000 0000
Step 1 ->B12:1         0000 0000 0000 0001         B12:17      0000 0000 0000 0000
Step 2 ->B12:2         0000 0000 0000 0010         B12:18      0000 0000 0000 0000
Step 3 ->B12:3         0000 0000 0000 0100         B12:19      0000 0000 0000 0000
Step 4 ->B12:4         0000 0000 0000 1000         B12:20      0000 0000 0000 0000
Step 5 ->B12:5         0000 0000 0001 0000         B12:21      0000 0000 0000 0000
Step 6 ->B12:6         0000 0000 0010 0000         B12:22      0000 0000 0000 0000
Step 7 ->B12:7         0000 0000 0100 0000         B12:23      0000 0000 0000 0000
Step 8 ->B12:8         0000 0000 1000 0000         B12:24      0000 0000 0000 0000
Step 9 ->B12:9         0000 0001 0000 0000
Step 10->B12:10        0000 0010 0000 0000
Step 11->B12:11        0000 0100 0000 0000
Step 12->B12:12        0000 1000 0000 0000
Step 13->B12:13        0001 0000 0000 0000
Step 14->B12:14        0010 0000 0000 0000
Step 15->B12:15        0100 0000 0000 0000


Press a key or enter value
B12/271 =
offline           no forces        binary data           decimal addr   File SEQ_ONE
CHANGE                               SPECIFY              NEXT   PREV
   RADIX                             ADDRESS              FILE   FILE
     F1                                 F5                 F7     F8
```

Figure 37-2   Sequencer bit file B:12.

1. _____   Select F9, *Data Monitor*.
             Enter the data table address > *B12*.
2. _____   *Enter*.
3. _____   Enter data into the table in binary radix.
4. _____   Use arrow keys to move cursor to B12/16. This is word one, address B12:1 or address B12/16 (bit file 12, bit 16). Remember that word zero of a sequencer file is step zero, and must not be used as a machine sequence step.
5. _____   Enter *1*.
6. _____   *Enter*.
             This is step one of our sequence where output zero will energize.
7. _____   Cursor down to B12/33.
8. _____   Enter *1*.
9. _____   *Enter*. This is set two of the sequence. Bit one will energize output one on output module address O:2/1.
10. _____   Move cursor to B12/50.
11. _____   Enter *1*.
12. _____   *Enter*. This is sequencer step 3.
13. _____   Complete entering data as illustrated in Figure 37-3 on the next page.

You can key in each bit in each 16-bit word, or convert each 16-bit data word to Hex. Changing the data tables radix to hex will allow you to key in four numbers per bit file element rather than sixteen ones or zeros.

```
  address        15      data       0      address        15      data       0
B12:0          0000 0000 0000 0000      B12:16         1000 0000 0000 0000
B12:1          0000 0000 0000 0001      B12:17         0000 0000 0000 0000
B12:2          0000 0000 0000 0010      B12:18         0000 0000 0000 0000
B12:3          0000 0000 0000 0100      B12:19         0000 0000 0000 0000
B12:4          0000 0000 0000 1000      B12:20         0000 0000 0000 0000
B12:5          0000 0000 0001 0000      B12:21         0000 0000 0000 0000
B12:6          0000 0000 0010 0000      B12:22         0000 0000 0000 0000
B12:7          0000 0000 0100 0000      B12:23         0000 0000 0000 0000
B12:8          0000 0000 1000 0000      B12:24         0000 0000 0000 0000
B12:9          0000 0001 0000 0000
B12:10         0000 0010 0000 0000
B12:11         0000 0100 0000 0000
B12:12         0000 1000 0000 0000
B12:13         0001 0000 0000 0000
B12:14         0010 0000 0000 0000
B12:15         0100 0000 0000 0000


Press a key or enter value
B12/271 =
offline            no forces           binary data    decimal addr    File SEQ_ONE
CHANGE                                 SPECIFY        NEXT   PREV
  RADIX                                ADDRESS        FILE   FILE
   F1                                    F5            F7     F8
```

Figure 37-3   Completed sequencer file B:12.

CHANGE THE DATA TABLE RADIX

To change the radix of the data table:

1. _____ Press F1, *Change Radix*.
2. _____ Press F4, *HEX / BCD*. This will allow you to enter or view data in either hexadecimal or BCD format. Use the arrow keys to move between bit file words. Data will be entered in Hex format. Enter data where the cursor is blinking next to the bit file address. As an example, where it says "Press a key or enter a value B12:2 = ," enter the value 2E.
3. _____ Press *Enter*. The hex value should be displayed in the data file in the proper address location.
4. _____ When completed entering data, press *Escape* to return to the ladder program.
5. _____ *Save and Go on Line*.
6. _____ F8, *Data Monitor*.
7. _____ Cycle sequencer instruction's input instruction. On-Off cycling will cause the outputs to sequence as the sequencer steps through its programmed sequence.
8. _____ Notice that when you cycle the input instruction the sequencer steps to the next step in the sequence.

QUESTIONS

1. List the SQO instruction parameters._____

2. Explain the usage of the SQO instruction._____

_____

3. Explain why the mask is used with the SQO instruction._____

_____

4. What is the purpose of the control parameter?_____

5. Why does the control parameter have an address associated with it?_____

_____

6. Typically the control address begins as R6. Explain the significance of the R and the 6._____

_____

7. Explain the file portion of the SQO instruction._____

_____

8. Explain the significance of the # in #B12:0._____

9. What mask would you use to control only the lower eight outputs on the output module in slot three?_____

10. What mask would you use to control only the upper eight outputs on the output module in slot three?_____

11. Explain how the sequencer instruction controls sixteen outputs on an output module when there is no output coil associated with the instruction._____

_____

12. Explain the significance of the *length* portion of the SQO instruction._____

_____

13. What does the value in the *position* portion of the SQO instruction tell you?_____

_____

14. In Figure 37-4, illustrate the control word structure. Identify all parts and their significance.

0	
1	
2	

Figure 37-4

15. Assume you have an output module in slot 2 of your SLC 500 chassis. You need a 12-step sequencer. Inputs O:2/0 and O:2/1 are currently used to control field devices not associated with the sequence operation. Since this is the only output module with output points available, can you use this module for the sequence and still use O:2/0 and O:2/1 as separate outputs?

_____

16. If you answered yes to question 15, explain how._____

_____

38

SEQUENCER PROGRAM EDITING

In this exercise we will edit our sequencer program for Exercise 37 to have timer four input a two-second pulse into the sequencer. Now each sequence step will increment every two seconds automatically. Figure 38-1 illustrates the completed program.

```
Rung 2:0
|   I:1   T4:0                                            +TON--------------+     |
|--] [---]/[-----------------------------------------+TIMER ON DELAY   +-(EN)- |
|    1      DN                                        |Timer        T4:0+-(DN) |
|                                                     |Time Base     1.0|       |
|                                                     |Preset          2|       |
|                                                     |Accum           0|       |
|                                                     +-----------------+       |
```

```
Rung 2:1
Each time T4:0/DN is energized by timer T4:0, the sequencer instruction (SQO)
will increment to the next position, or step in the sequence.  When all of the
steps have been stepped through (the length) the sequence will begin at the
first step and step through the sequence again.
|   T4:0                                               +SQO--------------+     |
|--] [------------------------------------------------+SEQUENCER OUTPUT  +-(EN)- |
|    DN                                                |File        #B12:0+-(DN) |
|                                                      |Mask          FFFF|      |
|                                                      |Dest         O:2.0|      |
|                                                      |Control       R6:1|      |
|                                                      |Length          16|      |
|                                                      |Position         2|      |
|                                                      +-----------------+       |
```

Figure 38-1 *SEQ_TWO* ladder program.

1. _____ Edit *SEQ_ONE* program.
2. _____ *Insert Rung.*
3. _____ *Insert Instruction.*
4. _____ F1, *Bit.*

5. _____ F1, *XIC*.

Enter bit address > *I:1/1*.

6. _____ Enter the XIO instruction.

7. _____ Select F2, *Timer Counter*.

8. _____ F1, *TON*.

9. _____ Enter timer information for T4:0, 1 second time base, 2 second preset.

10. _____ Press *Enter* when finished entering timer data.

11. _____ *Escape.*

12. _____ *Accept Rung.*

13. _____ Cursor to the sequencer rung I:1/0.

14. _____ Modify Rung and the instruction to an XIC addressed as T4:0/DN.

17. _____ *Accept Rung.*

18. _____ *Save/ Go On Line.*

20. _____ Put PLC into program mode.

21. _____ Verify program operation.

22. _____ Save this exercise to floppy. Title it *SEQ_TWO*.

APPENDIX

COMMONLY USED PROCEDURES

HOW TO CREATE A NEW DATA FILE

As an example you want to create a new integer file, N12. The new file N12 needs to be 100 words long. The procedure is listed below.

1. _____ From the main screen, press F3, *Offline Programming and Documentation*.
2. _____ F10, *Memory Map*.
3. _____ F6, *Create Data Table File*.
        Enter address to create: *N12:99* (100 words, 0 through 99)
4. _____ *Enter.*
5. _____ Press *Escape* to return to the program directory screen.
6. _____ Press F2, *Save*.
7. _____ To the message, "File already exists. Overwrite file?" Select F8, *Yes*.
8. _____ Accept the defaults for: a. Index across files; b. Force protection; c. Future access.

EDIT SLC 500 I/O CONFIGURATION

In the event you need to edit an SLC 500 or MicroLogix 1000's I/O configuration, follow the following steps:

1. _____ From the main screen, F3, *Off line Programming and Documentation*.
2. _____ F1, *Processor Functions*.
3. _____ F1, *Change Processor*.
4. _____ Select F5 to modify racks or I/O. The following selections will become available:
        F1, *Read I/O Configuration* (only in model 5/03 and above).
        F2, *Configure on line communications*.
        F4, *Modify current rack or racks*.

F5, *Modify highlighted rack I/O slot's module.*
F6, *Delete highlighted rack I/O slot's module.*
F7, *Undelete the highlighted rack I/O slot's module.*
F8, *Exit I/O configuration.*
F9, *Specialty module I/O configuration.*

5. _____ When completed, select *Exit.*
6. _____ Select F8, *Save & Exit.*

ON-LINE CONFIGURATION

This sets up communication parameters so your personal computer can communicate with your SLC 500.

1. _____ From the main menu screen select F2, *On Line Config.:*
2. _____ Typical defaults should be as follows:

F1	Port	COM 1
F2	Current Device	1747 PIC (DH-485)
F3	Baud Rate	19200
F4	Terminal Address	0
F5	Processor Address	1
F6	Max Node Address	31
F9	Save To File	

3. _____ On Line Configuration can also be accessed from the F5 *Who* screen:
   1. _____ Select F5, *Who.*
   2. _____ Select F4, *Who Listen.*
   3. _____ From *Who Listen* select:
      F2, *On Line Configuration.*
4. _____ To change your terminal's address select F5 and type in the new address, which should display on WHO.

SAVING YOUR COMPLETED LADDER PROGRAM WITH THE DESIRED NAME TO FLOPPY DISK

When completed development of your ladder program, save your work to your student floppy disk for future editing and reference.

1. _____ Begin at the main menu.
2. _____ Press F7, *File Options.*
   File Options will provide you with the following selections:

F3	Rename
F4	Copy
F5	Delete
F7	Copy To Disk
F8	Copy From Disk

3. _____ Select F7, *Copy To Disk* to copy your hard drive program to a floppy.
   You will see the following selections:

F1	Processor Memory File
F3	Comments And Symbols
F4	Documentation Reports
F5	All Of The Above

4. _____ Select F5, *All Of The Above.*

5. _____ Use arrow keys to select the file to be copied.
6. _____ Press F3 to select this file.
7. _____ Press F1, *Begin Operation.* File will be saved to floppy.
8. _____ Press *Escape* to return to file operations menu.
9. _____ Press *Escape* to return the main menu.

TO CHANGE PROCESSOR DH-485 NODE ADDRESS

To change your processor's DH-485 node address:

1. _____ From the main menu screen press F5, *WHO.*
2. _____ Press F5, *Who Active.*
3. _____ Cursor to current processor.
4. _____ Press F8 and type in new address.
5. _____ Press *ENTER.*
6. _____ Cycle power to your SLC 500 PLC. PLC should come up as new node address when power comes back up. The Who Active screen should also show the new node address.

DEVELOPING A NEW PROGRAM

To develop a new program, start with the main menu and:

1. _____ Select F3, *Off Line P.G./DOC.*
2. _____ Select F4, *Change File.*
3. _____ Select F6, *Create File.* (This will create a new processor file)
   Type in your new program name.
   Press *Enter* key.
4. _____ Select processor, and press F2.
5. _____ Select F5, *Configure I/O.*
   A. If you have a 5/03 processor and above, along with its associated software, press F1, *Read Configuration,* and I/O will be configured when you press F8, *YES.*
   B. For other processors:
   Select F4, *Modify Rack.*
   1. Select F1, *Rack 1.*
   2. Use arrow key to select proper rack.
   3. Press *Enter* key.
   Press F5, *Modify Slot.*
   1. Use arrow keys to find correct module in slot 1.
   2. Press F2 to select.
   3. Press F5, *Modify Slot,* to choose module in slot 2.
   4. Use arrow keys to select correct module.
   5. Press F2 to select.
   6. Complete this sequence until all slots have been selected.
   7. Press F8 to exit when completed.
   8. Press *Save and Exit.*
   9. Press F8, *Accept.*
   10. Press F2, *Save.*
   11. The message "File already exists, overwrite file?"
   12. Press F8. When the stars disappear from size column the processor has completed creating the default data files for this processor file.

DEVELOPING A NEW LADDER

1. _____ Press F8, *Monitor File*.
2. _____ Press F10, *Edit*.
3. _____ Press F4, *Insert Rung*.
4. _____ Press F4, *Insert Instruction*.
5. _____ Select the desired instruction.
6. _____ Enter the instructions address.
7. _____ Enter the next instruction.
8. _____ When all of the rung's instructions have been entered, press *Escape*.
9. _____ Press F10 to accept the rung.
10. _____ Insert or append the next rung as desired.